C000232123

Spon's Estimating (.ɔ
Minor Works, Alteraᴛᴉᴏns and Repairs
to Fire, Flood, Gale and Theft Damage

Other books by Bryan Spain

also available from Spon Press '

Spon's Estimating Costs Guide to Plumbing and Heating
(2004 edition)
Pb: 0-415-31855-6

Spon's Estimating Costs Guide to Electrical Works
(2004 edition)
Pb: 0-415-31853-X

*Spon's House Improvement Price Book: house extensions, storm
damage work, alterations, loft conversions and insulation* (2003
edition)
Pb: 0-415-30938-7

Spon's First Stage Estimating Price Book (2000 edition)
Pb: 0-415-23436-0

Spon's Construction Resource Handbook (1998 edition)
Hb: 0-419-23680-5

Information and ordering details

For price availability and ordering visit our website **www.sponpress.com**

Alternatively our books are available from all good bookshops.

Spon's Estimating Costs Guide to Minor Works, Alterations and Repairs to Fire, Flood, Gale and Theft Damage

Second edition

Bryan Spain

Spon's Contractors' Handbooks

Spon Press
Taylor & Francis Group

LONDON AND NEW YORK

First published 1999 by Spon Press
Second edition 2004
11 New Fetter Lane, London EC4P 4EE

Simultaneously published in the USA and Canada
by Spon Press
29 West 35th Street, New York, NY 10001

Spon Press is an imprint of the Taylor & Francis Group

Publisher's Note
This book has been prepared from camera-ready copy supplied by the author.

Printed and bound in Great Britain by TJ International Ltd, Padstow, Cornwall.

British Library Cataloguing in Publication Data
A catalogue record for this book is available from the British Library

Library of Congress Cataloging in Publication Data
A catalog record for this book has been requested

ISBN 0-415-31854-8

Contents

Masonry

Carpentry and joinery

Metalwork

Roofing

Asphalt work

Floor, wall and ceiling finishings

Plumbing and heating

Glazing

Painting and wallpapering

Part Three: Approximate estimating

Part Four: Plant and tool hire

Part Five: General construction data

Part Six: Business matters

Preface

The need for accurate estimates that are prepared quickly is vital for all contractors operating in the domestic construction market. Most contractors have the skills necessary to carry out the work together with the capacity for dealing with the setbacks that are part of the normal construction process. What they do not have is enough time to complete the many tasks that must be carried out in order to trade profitably. This book aims to help contractors to prepare their estimates by providing thousands of rates for small building work. If used sensibly, the book can save valuable time for contractors in the preparation of their bids and can also play a part in making sure that value for money is received from sub-contractors.

I have received a great deal of support in the research necessary for this type of book and I am grateful to those individuals and firms who have provided the cost data and other information. In particular, I am indebted to Mark Loughrey of Youds, Ellison & Co., Chartered Accountants of Hoylake (tel: 0151-632 3298 or www.yesl.uk.com), who are specialists in advising small construction businesses. Their research for the information in the business section is based on tax legislation in force in December 2003.

Although every care has been taken in the preparation of the book, neither the publishers nor I can accept any responsibility for the use of the information provided by any firm or individual. Finally, I would welcome any constructive criticism of the book's contents and suggestions that could be incorporated into future editions.

Bryan Spain
www.costofdiy.com
December 2003

Introduction

This edition of Spon's Minor Works, Refurbishment and Repairs book follows the layout, style and contents of previous editions. In the last edition, a new section was introduced to cover damage caused by fire, flood, gales and theft. This was well received and is included again at the request of those contractors and others involved in this section of construction market.

The contents of this book cover unit rates, general advice on business matters and other information useful to those involved in the commissioning and construction of building work. The unit rates section presents analytical rates for work up to about £50,000 in value. The business section covers advice on starting and running a business together with information on taxation and VAT matters.

Materials

Small contractors are not able to purchase materials in large quantities so it has been assumed that they will probably not receive the benefits of the discounts available to larger contractor. An average of 10% to 15% discount has been allowed on normal trade prices.

Labour

The hourly labour rates for craftsmen and general operatives are based upon the current wage awards promulgated by the Building and Allied Trades Joint Industrial Council (BATJIC) and the Joint Industry Board for Plumbing Mechanical Engineering Services (JIBMES).

General operative	£9.80
Craftsman	£12.00
Advanced plumber	£14.50

These rates include provision for NIC Employers' contribution CITB levy, insurances, public and annual holidays, severance pay and tool allowances where appropriate.

Overheads and profit

This has been set at 15% for all grades of work and is deemed to cover
head office and site overheads including:

- heating
- lighting
- rent
- rates
- telephones
- secretarial services
- insurances
- finance charges
- transport
- small tools
- ladders
- scaffolding etc.

Standard Method of Measurement/trades link

The contents of this book are presented under trade headings and the following table provides a link to the Standard Method of Measurement (SMM).

Demolition, excavation and filling

> C10 Demolishing structures
> D20 Excavation and filling

Concrete work

> E10 In situ concrete
> E20 Formwork for in situ concrete
> E30 Reinforcement for in situ concrete
> E40 Designed joints for in situ concrete
> E41 Worked finished to situ concrete
> E50 Precast concrete units

Brickwork and blockwork

> F10 Brick/block walling
> F30 Accessories for brick/block walling

Masonry

> F20 Natural stone rubble walling
> F30 Accessories for stone walling

Carpentry and joinery

> G20 Carpentry/timber framing/first fixing
> G32 Edge supported woodwool decking
> K20 Timber board flooring
> L10 Timber windows
> L20 Timber doors
> L30 Timber stairs
> N11 Domestic fittings

Carpentry and joinery (cont'd)
> Y50 Thermal insulation
> P21 Ironmongery

Metalwork

> L31 Balustrades

Roofing

> H60 Clay/concrete roof tiling
> H61 Fibre-cement slating
> H62 Natural slating
> H63 Reconstructed stone slating
> H71 Lead sheet coverings
> H73 Copper sheet coverings
> J41 Built-up roofing

Asphalt work

> J20 Mastic asphalt tanking
> M11Mastic asphalt flooring
> J21 Mastic asphalt roofing

Floor, wall and ceiling finishes

> M10 Sand cement/granolithic screeds
> M20 Plastered coatings
> M30 Metal mesh lathing
> K10 Plasterboard dry lining
> M40 Quarry/ceramic tiling
> M50 Plastic/lino tiling

Plumbing and heating

> N13 Sanitary fittings
> R10 Rainwater pipes and gutters
> R11 Foul drainage above ground
> S10 Piped supply systems
> Y21Water tank/cisterns
> Y23 Storage cylinders
> S41 Fuel oil storage
> T31 Low temperature hot water heating

Glazing

> L40 General glazing

Painting and wallpapering

> M60 Painting/clear finishing
> M52 Decorative papers
> Internal metalwork
>
> Wallpapering

External works

> R12 Drainage below ground
> Q40 Fencing
> Q10 Concrete kerbs and edgings
> Q20 Hardcore/granular sub-bases
> Q25 Slab/brick pavings

Alterations and repairs

> C30 Shoring
> D50 Underpinning
> C10 Demolishing structures
> C20 Spot items

Part One

UNIT RATES

Demolition, excavation and filling

Concrete work

Brickwork and blockwork

Masonry

Carpentry and joinery

Metalwork

Roofing

Asphalt work

Floor, wall and ceiling finishings

Plumbing and heating

Glazing

Painting and wallpapering

External works

Alterations and repairs

Spot items

	Unit	Hours	Hours £	Plant £	O & P £	Total £

DEMOLITION, EXCAVATION AND FILLING

Demolition

Demolish ground level existing brick buildings

single storey, detached	m3	1.60	15.68	0.00	2.35	18.03
single storey, attached	m3	1.70	16.66	0.00	2.50	19.16
two storey, attached	m3	1.80	17.64	0.00	2.65	20.29

Break up debris and arisings, load into skips or lorries

unreinforced concrete slabs						
100mm thick	m2	0.80	7.84	0.50	1.25	9.59
150mm thick	m2	1.00	9.80	0.75	1.58	12.13
200mm thick	m2	1.40	13.72	1.00	2.21	16.93
reinforced concrete slabs						
100mm thick	m2	1.40	13.72	1.00	2.21	16.93
150mm thick	m2	2.00	19.60	1.25	3.13	23.98
200mm thick	m2	2.60	25.48	1.50	4.05	31.03
reinforced suspended concrete slabs						
100mm thick	m2	2.80	27.44	1.00	4.27	32.71
150mm thick	m2	3.20	31.36	1.25	4.89	37.50
200mm thick	m2	3.40	33.32	1.50	5.22	40.04
reinforced concrete beams						
200 × 300mm	m	1.25	12.25	0.20	1.87	14.32
250 × 350mm	m	1.50	14.70	0.30	2.25	17.25
300 × 400mm	m	1.75	17.15	0.40	2.63	20.18
reinforced concrete columns						
200 × 300mm	m	1.30	12.74	0.20	1.94	14.88
250 × 350mm	m	1.60	15.68	0.30	2.40	18.38
300 × 400mm	m	1.90	18.62	0.40	2.85	21.87

	Unit	Hours	Hours £	Plant £	O & P £	Total £
Demolish brick walls in cement mortar						
half brick thick	m2	0.80	7.84	0.00	1.18	9.02
one brick thick	m2	1.40	13.72	0.00	2.06	15.78
one and a half brick thick	m2	1.90	18.62	0.00	2.79	21.41
two brick thick	m2	2.70	26.46	0.00	3.97	30.43
Demolish brick walls in cement lime mortar						
half brick thick	m2	0.70	6.86	0.00	1.03	7.89
one brick thick	m2	1.30	12.74	0.00	1.91	14.65
one and a half brick thick	m2	1.80	17.64	0.00	2.65	20.29
two brick thick	m2	2.50	24.50	0.00	3.68	28.18
Carefully take down brick walls in cement lime mortar, clean bricks and lay aside for re-use						
half brick thick	m2	1.00	9.80	0.00	1.47	11.27
one brick thick	m2	1.70	16.66	0.00	2.50	19.16
one and a half brick thick	m2	2.20	21.56	0.00	3.23	24.79
two brick thick	m2	3.20	31.36	0.00	4.70	36.06
Take out fireplace and surround, load into skips or lorries						
1500 × 1200mm high	nr	1.50	14.70	0.00	2.21	16.91
1500 × 1500mm high	nr	1.75	17.15	0.00	2.57	19.72
1800 × 1500mm high	nr	2.00	19.60	0.00	2.94	22.54
Demolish blockwork partitions						
75mm thick	m2	0.80	7.84	0.00	1.18	9.02
100mm thick	m2	0.90	8.82	0.00	1.32	10.14
115mm thick	m2	1.00	9.80	0.00	1.47	11.27
140mm thick	m2	1.10	10.78	0.00	1.62	12.40
190mm thick	m2	1.20	11.76	0.00	1.76	13.52
215mm thick	m2	1.40	13.72	0.00	2.06	15.78

	Unit	Hours	Hours £	Plant £	O & P £	Total £

Excavation

Excavate by hand

topsoil
150mm thick	m2	0.35	3.43	0.00	0.51	3.94
200mm thick	m2	0.45	4.41	0.00	0.66	5.07
250mm thick	m2	0.60	5.88	0.00	0.88	6.76

to reduce levels, depth not exceeding
0.25m	m3	2.20	21.56	0.00	3.23	24.79
1.00m	m3	2.40	23.52	0.00	3.53	27.05
1.50m	m3	2.80	27.44	0.00	4.12	31.56
2.00m	m3	3.20	31.36	0.00	4.70	36.06

to trenches 450mm wide, depth not exceeding
0.25m	m3	2.50	24.50	0.00	3.68	28.18
1.00m	m3	2.70	26.46	0.00	3.97	30.43
1.50m	m3	3.00	29.40	0.00	4.41	33.81
2.00m	m3	3.40	33.32	0.00	5.00	38.32

to pits, depth not exceeding
0.25m	m3	2.50	24.50	0.00	3.68	28.18
1.00m	m3	2.70	26.46	0.00	3.97	30.43
1.50m	m3	3.10	30.38	0.00	4.56	34.94
2.00m	m3	3.50	34.30	0.00	5.15	39.45

Extra for excavating through

rock	m3	10.00	98.00	0.00	14.70	112.70
concrete	m3	8.00	78.40	0.00	11.76	90.16
reinforced concrete	m3	9.00	88.20	0.00	13.23	101.43
brickwork	m3	6.00	58.80	0.00	8.82	67.62

	Unit	Hours	Hours £	Plant £	O & P £	Total £
Excavate by machine						
topsoil						
150mm thick	m2	0.04	0.39	1.20	0.24	1.83
200mm thick	m2	0.05	0.49	1.30	0.27	2.06
250mm thick	m2	0.06	0.59	1.40	0.30	2.29
to reduce levels, depth not exceeding						
0.25m	m3	0.10	0.98	1.40	0.36	2.74
1.00m	m3	0.08	0.78	1.20	0.30	2.28
1.50m	m3	0.10	0.98	1.40	0.36	2.74
2.00m	m3	0.12	1.18	1.60	0.42	3.19
to trenches 450mm wide, depth not exceeding						
0.25m	m3	0.35	3.43	4.00	1.11	8.54
1.00m	m3	0.30	2.94	3.80	1.01	7.75
1.50m	m3	0.35	3.43	4.00	1.11	8.54
2.00m	m3	0.40	3.92	4.50	1.26	9.68
to pits, depth not exceeding						
0.25m	m3	0.40	3.92	4.50	1.26	9.68
1.00m	m3	0.35	3.43	4.20	1.14	8.77
1.50m	m3	0.40	3.92	4.50	1.26	9.68
2.00m	m3	0.45	4.41	4.80	1.38	10.59
Extra for excavating through						
rock	m3	5.00	49.00	16.00	9.75	74.75
concrete	m3	3.75	36.75	12.50	7.39	56.64
reinforced concrete	m3	4.00	39.20	14.30	8.03	61.53
brickwork	m3	3.80	37.24	11.50	7.31	56.05

	Unit	Hours	Hours £	Mat'ls £	O & P £	Total £

Earthwork support

Earthwork support not exceeding
2m between opposite faces,
depth not exceeding 1m

	Unit	Hours	Hours £	Mat'ls £	O & P £	Total £
firm ground	m2	0.15	1.47	1.20	0.40	3.07
loose ground	m2	0.85	8.33	2.40	1.61	12.34
sand	m2	1.10	10.78	2.95	2.06	15.79

Earthwork support not exceeding
2m between opposite faces,
depth not exceeding 2m

	Unit	Hours	Hours £	Mat'ls £	O & P £	Total £
firm ground	m2	0.18	1.76	1.20	0.44	3.41
loose ground	m2	0.90	8.82	2.40	1.68	12.90
sand	m2	1.20	11.76	2.95	2.21	16.92

Earthwork support not exceeding
2m between opposite faces,
depth not exceeding 4m

	Unit	Hours	Hours £	Mat'ls £	O & P £	Total £
firm ground	m2	0.22	2.16	1.20	0.50	3.86
loose ground	m2	0.95	9.31	2.40	1.76	13.47
sand	m2	1.30	12.74	2.95	2.35	18.04

Disposal by hand

Load surplus excavated material
into barrows, wheel and deposit
in temporary spoil heaps, average
distance

	Unit	Hours	Hours £	Mat'ls £	O & P £	Total £
15m	m3	1.25	12.25	0.00	1.84	14.09
25m	m3	1.45	14.21	0.00	2.13	16.34
50m	m3	1.70	16.66	0.00	2.50	19.16

	Unit	Hours	Hours £	Plant £	O & P £	Total £
Load surplus excavated material into barrows, wheel and spread over site, average distance						
15m	m3	1.65	16.17	0.00	2.43	18.60
25m	m3	1.85	18.13	0.00	2.72	20.85
50m	m3	2.00	19.60	0.00	2.94	22.54
Load surplus excavated material into barrows, wheel and deposit in skips or lorries, average distance						
15m	m3	1.20	11.76	0.00	1.76	13.52
25m	m3	1.40	13.72	0.00	2.06	15.78
50m	m3	1.65	16.17	0.00	2.43	18.60

Disposal by machine

	Unit	Hours	Hours £	Plant £	O & P £	Total £
Load and deposit in temporary spoil heaps, average distance						
15m	m3	0.10	0.98	0.90	0.28	2.16
25m	m3	0.15	1.47	1.10	0.39	2.96
50m	m3	0.25	2.45	1.60	0.61	4.66
Load and spread over site, average distance						
15m	m3	0.15	1.47	1.10	0.39	2.96
25m	m3	0.18	1.76	1.25	0.45	3.47
50m	m3	0.24	2.35	1.40	0.56	4.31
Load and deposit in skips or lorries, average distance						
15m	m3	0.08	0.78	0.90	0.25	1.94
25m	m3	0.12	1.18	1.10	0.34	2.62
50m	m3	0.22	2.16	1.60	0.56	4.32

	Unit	Hours	Hours £	Mat'ls £	O & P £	Total £
Filling by hand						
Surplus excavated material deposited and compacting in layers						
over 250mm thick	m3	1.20	11.76	2.66	2.16	16.58
100mm thick	m2	0.20	1.96	0.40	0.35	2.71
150mm thick	m2	0.35	3.43	0.50	0.59	4.52
200mm thick	m2	0.50	4.90	0.60	0.83	6.33
Imported sand deposited and compacting in layers						
over 250mm thick	m3	1.20	11.76	28.80	6.08	46.64
100mm thick	m2	0.20	1.96	2.80	0.71	5.47
150mm thick	m2	0.35	3.43	4.20	1.14	8.77
200mm thick	m2	0.50	4.90	5.60	1.58	12.08
Imported hardcore deposited and compacting in layers						
over 250mm thick	m3	1.20	11.76	26.60	5.75	44.11
100mm thick	m2	0.20	1.96	2.66	0.69	5.31
150mm thick	m2	0.35	3.43	3.99	1.11	8.53
200mm thick	m2	0.50	4.90	5.32	1.53	11.75
Filling by machine						
Surplus excavated material deposited and compacting in layers						
over 250mm thick	m3	0.40	3.92	2.66	0.99	7.57
100mm thick	m2	0.06	0.59	0.40	0.15	1.14
150mm thick	m2	0.08	0.78	0.50	0.19	1.48
200mm thick	m2	0.10	0.98	0.60	0.24	1.82

	Unit	Hours	Hours £	Mat'ls £	O & P £	Total £
Imported sand deposited and compacting in layers						
over 250mm thick	m3	0.40	3.92	28.80	4.91	37.63
100mm thick	m2	0.06	0.59	2.80	0.51	3.90
150mm thick	m2	0.08	0.78	4.20	0.75	5.73
200mm thick	m2	0.10	0.98	5.60	0.99	7.57
Imported hardcore deposited and compacting in layers						
over 250mm thick	m3	0.40	3.92	26.60	4.58	35.10
100mm thick	m2	0.06	0.59	2.66	0.49	3.74
150mm thick	m2	0.08	0.78	3.99	0.72	5.49
200mm thick	m2	0.10	0.98	5.32	0.95	7.25

Surface treatments

	Unit	Hours	Hours £	Mat'ls £	O & P £	Total £
Level and compact bottom of excavation with vibrating roller	m2	0.10	0.98	3.65	0.69	5.32
Blind filling surfaces with sand 50mm thick	m2	0.12	1.18	1.40	0.39	2.96

	Unit	Hours	Hours £	Mat'ls £	O & P £	Total £

CONCRETE WORK

Ready-mixed concrete 1:3:6
40mm aggregate

Foundations in trenches, thickness

150 to 300mm	m3	1.85	18.13	74.73	13.93	106.79
300 to 450mm	m3	1.35	13.23	74.73	13.19	101.15
over 450mm	m3	1.15	11.27	74.73	12.90	98.90

Isolated column bases

150 to 300mm	m3	2.00	19.60	74.73	14.15	108.48
300 to 450mm	m3	1.50	14.70	74.73	13.41	102.84
over 450mm	m3	1.30	12.74	74.73	13.12	100.59

Filling to cavity walls	m3	3.95	38.71	74.73	17.02	130.46

Ready-mixed concrete 1:2:4
20mm aggregate

Beds

150 to 300mm	m3	2.40	23.52	79.81	15.50	118.83
300 to 450mm	m3	2.10	20.58	79.81	15.06	115.45
over 450mm	m3	1.80	17.64	79.81	14.62	112.07

Slabs

150 to 300mm	m3	3.20	31.36	79.81	16.68	127.85
300 to 450mm	m3	2.85	27.93	79.81	16.16	123.90

Walls

150 to 300mm	m3	4.25	41.65	79.81	18.22	139.68
300 to 450mm	m3	3.95	38.71	79.81	17.78	136.30

	Unit	Hours	Hours £	Mat'ls £	O & P £	Total £
Casings						
isolated beams	m3	5.00	49.00	79.81	19.32	148.13
isolated deep beams	m3	4.50	44.10	79.81	18.59	142.50
attached deep beams	m3	5.00	49.00	79.81	19.32	148.13
columns	m3	5.00	49.00	79.81	19.32	148.13
staircases	m3	6.00	58.80	79.81	20.79	159.40
Upstands and kerbs	m3	4.40	43.12	72.56	17.35	133.03

**Site-mixed concrete 1:3:6
40mm aggregate**

Foundations in trenches, thickness

	Unit	Hours	Hours £	Mat'ls £	O & P £	Total £
150 to 300mm	m3	2.85	27.93	68.37	14.45	110.75
300 to 450mm	m3	2.35	23.03	68.37	13.71	105.11
over 450mm	m3	2.15	21.07	68.37	13.42	102.86

Isolated column bases

	Unit	Hours	Hours £	Mat'ls £	O & P £	Total £
150 to 300mm	m3	3.00	29.40	68.37	14.67	112.44
300 to 450mm	m3	2.50	24.50	68.37	13.93	106.80
over 450mm	m3	2.30	22.54	68.37	13.64	104.55
Filling to cavity walls	m3	4.95	48.51	68.37	17.53	134.41

**Site-mixed concrete 1:2:4
20mm aggregate**

Beds

	Unit	Hours	Hours £	Mat'ls £	O & P £	Total £
150 to 300mm	m3	3.40	33.32	76.42	16.46	126.20
300 to 450mm	m3	3.10	30.38	76.42	16.02	122.82
over 450mm	m3	1.90	18.62	76.42	14.26	109.30

Slabs

	Unit	Hours	Hours £	Mat'ls £	O & P £	Total £
150 to 300mm	m3	4.20	41.16	76.42	17.64	135.22
300 to 450mm	m3	3.85	37.73	76.42	17.12	131.27

	Unit	Hours	Hours £	Mat'ls £	O & P £	Total £
Walls						
150 to 300mm	m3	5.25	51.45	76.42	19.18	147.05
300 to 450mm	m3	4.95	48.51	76.42	18.74	143.67
Casings						
isolated beams	m3	6.00	58.80	76.42	20.28	155.50
isolated deep beams	m3	5.50	53.90	76.42	19.55	149.87
attached deep beams	m3	6.00	58.80	76.42	20.28	155.50
columns	m3	6.00	58.80	76.42	20.28	155.50
staircases	m3	7.00	68.60	76.42	21.75	166.77
Upstands and kerbs	m3	5.50	53.90	76.42	19.55	149.87

Formwork

Plain vertical to sides of foundations

	Unit	Hours	Hours £	Mat'ls £	O & P £	Total £
over 1m	m2	2.30	22.54	8.58	4.67	35.79
not exceeding 250mm	m	0.75	7.35	3.01	1.55	11.91
250 to 500mm	m	1.25	12.25	5.21	2.62	20.08
500mm to 1m	m	1.70	16.66	8.58	3.79	29.03

Plain vertical to sides of
foundations, left in

	Unit	Hours	Hours £	Mat'ls £	O & P £	Total £
over 1m	m2	2.15	21.07	20.86	6.29	48.22
not exceeding 250mm	m	0.65	6.37	7.42	2.07	15.86
250 to 500mm	m	1.15	11.27	13.72	3.75	28.74
500mm to 1m	m	1.60	15.68	20.86	5.48	42.02

Plain vertical to sides of ground
beams and beds

	Unit	Hours	Hours £	Mat'ls £	O & P £	Total £
over 1m	m2	2.40	23.52	8.58	4.82	36.92
not exceeding 250mm	m	0.82	8.04	3.01	1.66	12.70
250 to 500mm	m	1.34	13.13	5.21	2.75	21.09
500mm to 1m	m	1.78	17.44	8.58	3.90	29.93

	Unit	Hours	Hours £	Mat'ls £	O & P £	Total £
Plain vertical to sides of ground beams and beds, left in						
over 1m	m2	2.25	22.05	20.86	6.44	49.35
not exceeding 250mm	m	0.75	7.35	7.42	2.22	16.99
250 to 500mm	m	1.25	12.25	13.72	3.90	29.87
500mm to 1m	m	1.70	16.66	20.86	5.63	43.15
Edges of suspended slabs						
not exceeding 250mm	m	0.95	9.31	3.01	1.85	14.17
250 to 500mm	m	1.40	13.72	5.21	2.84	21.77
Sides of upstands						
over 1m	m2	2.60	25.48	8.58	5.11	39.17
not exceeding 250mm	m	1.00	9.80	3.01	1.92	14.73
250 to 500mm	m	1.50	14.70	5.21	2.99	22.90
500mm to 1m	m	1.90	18.62	8.58	4.08	31.28
Steps in top surfaces						
not exceeding 250mm	m	0.75	7.35	3.01	1.55	11.91
250 to 500mm	m	1.25	12.25	5.21	2.62	20.08
500mm to 1m	m	1.70	16.66	8.58	3.79	29.03
Steps in soffits						
not exceeding 250mm	m	1.00	9.80	3.01	1.92	14.73
250 to 500mm	m	1.50	14.70	5.21	2.99	22.90
500mm to 1m	m	1.90	18.62	8.58	4.08	31.28
Machine bases and plinths						
over 1m	m2	2.30	22.54	8.58	4.67	35.79
not exceeding 250mm	m	0.75	7.35	3.01	1.55	11.91
250 to 500mm	m	1.25	12.25	5.21	2.62	20.08
500mm to 1m	m	1.70	16.66	8.58	3.79	29.03

	Unit	Hours	Hours £	Mat'ls £	O & P £	Total £
Horizontal to soffits of slabs, thickness						
not exceeding 200mm						
height to soffit, not exceeding						
1.5m	m2	2.20	21.56	8.58	4.52	34.66
height to soffit, 1.5 to 3m	m2	2.10	20.58	8.58	4.37	33.53
height to soffit, over 3m	m2	2.00	19.60	8.58	4.23	32.41
200 to 300mm						
height to soffit, not exceeding						
1.5m	m2	2.40	23.52	8.58	4.82	36.92
height to soffit, 1.5 to 3m	m2	2.30	22.54	8.58	4.67	35.79
height to soffit, over 3m	m2	2.20	21.56	8.58	4.52	34.66
300 to 400mm						
height to soffit, not exceeding						
1.5m	m2	2.60	25.48	8.58	5.11	39.17
height to soffit, 1.5 to 3m	m2	2.50	24.50	8.58	4.96	38.04
height to soffit, over 3m	m2	2.40	23.52	8.58	4.82	36.92
Horizontal to soffits of slabs, left in, thickness						
not exceeding 200mm						
height to soffit, not exceeding						
1.5m	m2	2.25	22.05	20.86	6.44	49.35
height to soffit, 1.5 to 3m	m2	2.20	21.56	20.86	6.36	48.78
height to soffit, over 3m	m2	2.10	20.58	20.86	6.22	47.66
200 to 300mm						
height to soffit, not exceeding						
1.5m	m2	2.35	23.03	20.86	6.58	50.47
height to soffit, 1.5 to 3m	m2	2.30	22.54	20.86	6.51	49.91
height to soffit, over 3m	m2	2.20	21.56	20.86	6.36	48.78

	Unit	Hours	Hours £	Mat'ls £	O & P £	Total £
Horizontal to soffits of slabs, left in, thickness (cont'd)						
300 to 400mm						
height to soffit, not exceeding						
1.5m	m2	2.45	24.01	20.86	6.73	51.60
height to soffit, 1.5 to 3m	m2	2.40	23.52	20.86	6.66	51.04
height to soffit, over 3m	m2	2.30	22.54	20.86	6.51	49.91
Walls						
vertical	m2	2.40	23.52	8.58	4.82	36.92
vertical interrupted	m2	2.50	24.50	8.58	4.96	38.04
Beams						
attached to slabs, height to soffit						
1.5 to 3m	m2	3.20	31.36	8.58	5.99	45.93
3 to 4.5m	m2	3.30	32.34	8.58	6.14	47.06
over 4.5m	m2	3.50	34.30	8.58	6.43	49.31
Columns						
attached to walls, height to soffit						
1.5 to 3m	m2	3.20	31.36	8.58	5.99	45.93
3 to 4.5m	m2	3.30	32.34	8.58	6.14	47.06
over 4.5m	m2	3.50	34.30	8.58	6.43	49.31
isolated, height to soffit						
1.5 to 3m	m2	3.40	33.32	8.58	6.29	48.19
3 to 4.5m	m2	3.50	34.30	8.58	6.43	49.31
over 4.5m	m2	3.60	35.28	8.58	6.58	50.44
Mortice in concrete for rag bolt, grout in cement mortar (1:3), depth						
50mm	nr	0.30	2.94	0.88	0.57	4.39
100mm	nr	0.35	3.43	1.87	0.80	6.10
150mm	nr	0.40	3.92	3.04	1.04	8.00

	Unit	Hours	Hours £	Mat'ls £	O & P £	Total £

Reinforcement

Plain round steel reinforcement
bars, straight or bent

	Unit	Hours	Hours £	Mat'ls £	O & P £	Total £
6mm diameter	m	0.02	0.20	0.16	0.05	0.41
8mm diameter	m	0.03	0.29	0.22	0.08	0.59
10mm diameter	m	0.04	0.39	0.32	0.11	0.82
12mm diameter	m	0.05	0.49	0.40	0.13	1.02
16mm diameter	m	0.06	0.59	0.63	0.18	1.40
20mm diameter	m	0.07	0.69	1.00	0.25	1.94
25mm diameter	m	0.08	0.78	1.62	0.36	2.76

High yield deformed steel
reinforcement bars, straight or bent

	Unit	Hours	Hours £	Mat'ls £	O & P £	Total £
8mm diameter	m	0.03	0.29	0.20	0.07	0.57
10mm diameter	m	0.04	0.39	0.29	0.10	0.78
12mm diameter	m	0.05	0.49	0.39	0.13	1.01
16mm diameter	m	0.06	0.59	0.41	0.15	1.15
20mm diameter	m	0.07	0.69	0.96	0.25	1.89
25mm diameter	m	0.08	0.78	1.57	0.35	2.71

Steel fabric reinforcement laid in
concrete beds

	Unit	Hours	Hours £	Mat'ls £	O & P £	Total £
A98 weighing 1.54kg per m2	m2	0.12	1.18	0.88	0.31	2.36
A142 weighing 2.22kg per m2	m2	0.14	1.37	1.03	0.36	2.76
A193 weighing 3.02kg per m2	m2	0.17	1.67	1.46	0.47	3.59
A252 weighing 3.95kg per m2	m2	0.19	1.86	1.79	0.55	4.20
B196 weighing 3.05kg per m2	m2	0.17	1.67	1.53	0.48	3.68
B283 weighing 3.73kg per m2	m2	0.18	1.76	1.74	0.53	4.03
B385 weighing 4.53kg per m2	m2	0.20	1.96	2.15	0.62	4.73
B503 weighing 5.93kg per m2	m2	0.22	2.16	2.78	0.74	5.68
B785 weighing 8.14kg per m2	m2	0.24	2.35	3.74	0.91	7.00
C283 weighing 2.61kg per m2	m2	0.15	1.47	1.43	0.44	3.34
C385 weighing 3.41kg per m2	m2	0.18	1.76	1.64	0.51	3.91
C503 weighing 4.34kg per m2	m2	0.20	1.96	1.88	0.58	4.42
C636 weighing 5.55kg per m2	m2	0.21	2.06	2.50	0.68	5.24
C785 weighing 6.72kg per m2	m2	0.23	2.25	3.30	0.83	6.39

	Unit	Hours	Hours £	Mat'ls £	O & P £	Total £

Joints

Impregnated fibreboard expansion
joint, width

12.5mm thick						
not exceeding 150mm thic	m	0.14	1.37	1.48	0.43	3.28
150 to 300mm thick	m	0.20	1.96	2.46	0.66	5.08
300 to 450mm thick	m	0.22	2.16	3.55	0.86	6.56
20mm thick						
not exceeding 150mm thic	m	0.16	1.57	2.17	0.56	4.30
150 to 300mm thick	m	0.22	2.16	3.27	0.81	6.24
300 to 450mm thick	m	0.24	2.35	4.89	1.09	8.33
25mm thick						
not exceeding 150mm thic	m	0.18	1.76	2.40	0.62	4.79
150 to 300mm thick	m	0.24	2.35	3.75	0.92	7.02
300 to 450mm thick	m	0.26	2.55	5.52	1.21	9.28

Concrete finishes

Treat surfaces of unset concrete

mechanical tamping	m2	0.07	0.69	0.00	0.10	0.79
power floating	m2	0.18	1.76	0.00	0.26	2.03
trowelling	m2	0.24	2.35	0.00	0.35	2.70
spade finish	m2	0.20	1.96	0.00	0.29	2.25
wood float finish	m2	0.16	1.57	0.00	0.24	1.80

Precast concrete

Reinforced concrete lintels
bedded in cement mortar (1:3)

100 × 75 × 900mm	nr	0.35	3.43	6.16	1.44	11.03
100 × 75 × 1000mm	nr	0.40	3.92	6.91	1.62	12.45
100 × 75 × 1200mm	nr	0.50	4.90	7.39	1.84	14.13
150 × 75 × 900mm	nr	0.45	4.41	6.64	1.66	12.71
150 × 75 × 1000mm	nr	0.45	4.41	7.51	1.79	13.71
150 × 75 × 1200mm	nr	0.55	5.39	9.00	2.16	16.55

	Unit	Hours	Hours £	Mat'ls £	O & P £	Total £
Precast concrete copings bedded in cement mortar (1:3)						
75 × 150mm	m	0.75	7.35	6.63	2.10	16.08
75 × 300mm	m	0.90	8.82	12.99	3.27	25.08

	Unit	Hours	Hours £	Mat'ls £	O & P £	Total £

BRICKWORK AND BLOCKWORK

Brickwork

Common bricks basic price £140
per thousand in cement mortar
(1:3)

walls

	Unit	Hours	Hours £	Mat'ls £	O & P £	Total £
half brick thick	m2	1.70	28.73	11.45	6.03	46.21
half brick thick overhand	m2	2.20	37.18	11.45	7.29	55.92
half brick thick against other work	m2	1.90	32.11	11.45	6.53	50.09
half brick thick curved	m2	2.30	38.87	11.45	7.55	57.87
one brick thick	m2	2.80	47.32	22.90	10.53	80.75
one brick thick curved	m2	3.40	57.46	22.90	12.05	92.41
one and a half brick thick	m2	3.50	59.15	34.35	14.03	107.53
two brick thick	m2	4.20	70.98	45.80	17.52	134.30
two brick thick battered	m2	4.80	81.12	45.80	19.04	145.96

walls, facework one side

	Unit	Hours	Hours £	Mat'ls £	O & P £	Total £
half brick thick	m2	1.80	30.42	11.45	6.28	48.15
half brick thick overhand	m2	2.30	38.87	11.45	7.55	57.87
half brick thick against other work	m2	2.00	33.80	11.45	6.79	52.04
half brick thick curved	m2	2.40	40.56	11.45	7.80	59.81
one brick thick	m2	2.90	49.01	22.90	10.79	82.70
one brick thick curved	m2	3.50	59.15	22.90	12.31	94.36
one and a half brick thick	m2	3.60	60.84	34.35	14.28	109.47
two brick thick	m2	4.30	72.67	45.80	17.77	136.24
two brick thick battered	m2	4.90	82.81	45.80	19.29	147.90

walls, facework both sides

	Unit	Hours	Hours £	Mat'ls £	O & P £	Total £
half brick thick	m2	1.90	32.11	11.45	6.53	50.09
half brick thick overhand	m2	2.40	40.56	11.45	7.80	59.81
half brick thick against other work	m2	2.10	35.49	11.45	7.04	53.98
half brick thick curved	m2	2.50	42.25	11.45	8.06	61.76
one brick thick	m2	3.00	50.70	22.90	11.04	84.64

	Unit	Hours	Hours £	Mat'ls £	O & P £	Total £
one brick thick curved	m2	3.60	60.84	22.90	12.56	96.30
one and a half brick thick	m2	3.70	62.53	34.35	14.53	111.41
two brick thick	m2	4.40	74.36	45.80	18.02	138.18
two brick thick battered	m2	5.00	84.50	45.80	19.55	149.85

Common bricks basic price £200
per thousand in cement mortar
(1:3)

walls
	Unit	Hours	Hours £	Mat'ls £	O & P £	Total £
half brick thick	m2	1.70	28.73	16.60	6.80	52.13
half brick thick overhand	m2	2.20	37.18	16.60	8.07	61.85
half brick thick against other work	m2	1.90	32.11	16.60	7.31	56.02
half brick thick curved	m2	2.30	38.87	16.60	8.32	63.79
one brick thick	m2	2.80	47.32	33.20	12.08	92.60
one brick thick curved	m2	3.40	57.46	33.20	13.60	104.26
one and a half brick thick	m2	3.50	59.15	49.80	16.34	125.29
two brick thick	m2	4.20	70.98	86.40	23.61	180.99
two brick thick battered	m2	4.80	81.12	45.80	19.04	145.96

walls, facework one side
	Unit	Hours	Hours £	Mat'ls £	O & P £	Total £
half brick thick	m2	1.90	32.11	16.60	7.31	56.02
half brick thick overhand	m2	2.40	40.56	16.60	8.57	65.73
half brick thick against other work	m2	2.10	35.49	16.60	7.81	59.90
half brick thick curved	m2	2.50	42.25	16.60	8.83	67.68
one brick thick	m2	3.00	50.70	33.20	12.59	96.49
one brick thick curved	m2	3.60	60.84	33.20	14.11	108.15
one and a half brick thick	m2	3.70	62.53	49.80	16.85	129.18
two brick thick	m2	4.40	74.36	86.40	24.11	184.87
two brick thick battered	m2	5.00	84.50	86.40	25.64	196.54

walls, facework both sides
	Unit	Hours	Hours £	Mat'ls £	O & P £	Total £
half brick thick	m2	1.90	32.11	16.60	7.31	56.02
half brick thick overhand	m2	2.40	40.56	16.60	8.57	65.73
half brick thick against other work	m2	2.10	35.49	16.60	7.81	59.90
half brick thick curved	m2	2.50	42.25	16.60	8.83	67.68
one brick thick	m2	3.00	50.70	33.20	12.59	96.49

	Unit	Hours	Hours £	Mat'ls £	O & P £	Total £
one brick thick curved	m2	3.60	60.84	33.20	14.11	108.15
one and a half brick thick	m2	3.70	62.53	49.80	16.85	129.18
two brick thick	m2	4.40	74.36	86.40	24.11	184.87
two brick thick battered	m2	5.00	84.50	86.40	25.64	196.54

Common bricks basic price £140
per thousand in gauged mortar
(1:1:6)

walls

	Unit	Hours	Hours £	Mat'ls £	O & P £	Total £
half brick thick	m2	1.70	28.73	11.30	6.00	46.03
half brick thick overhand	m2	2.20	37.18	11.30	7.27	55.75
half brick thick against other work	m2	1.90	32.11	11.30	6.51	49.92
half brick thick curved	m2	2.30	38.87	11.30	7.53	57.70
one brick thick	m2	2.80	47.32	22.60	10.49	80.41
one brick thick curved	m2	3.40	57.46	22.60	12.01	92.07
one and a half brick thick	m2	3.50	59.15	33.90	13.96	107.01
two brick thick	m2	4.20	70.98	45.20	17.43	133.61
two brick thick battered	m2	4.80	81.12	45.20	18.95	145.27

walls, facework one side

	Unit	Hours	Hours £	Mat'ls £	O & P £	Total £
half brick thick	m2	1.80	30.42	11.30	6.26	47.98
half brick thick overhand	m2	2.30	38.87	11.30	7.53	57.70
half brick thick against other work	m2	2.00	33.80	11.30	6.77	51.87
half brick thick curved	m2	2.40	40.56	11.30	7.78	59.64
one brick thick	m2	2.90	49.01	22.60	10.74	82.35
one brick thick curved	m2	3.50	59.15	22.60	12.26	94.01
one and a half brick thick	m2	3.60	60.84	33.90	14.21	108.95
two brick thick	m2	4.30	72.67	45.20	17.68	135.55
two brick thick battered	m2	4.90	82.81	45.20	19.20	147.21

walls, facework both sides

	Unit	Hours	Hours £	Mat'ls £	O & P £	Total £
half brick thick	m2	1.90	32.11	11.30	6.51	49.92
half brick thick overhand	m2	2.40	40.56	11.30	7.78	59.64
half brick thick against other work	m2	2.10	35.49	11.30	7.02	53.81
half brick thick curved	m2	2.50	42.25	11.30	8.03	61.58
one brick thick	m2	3.00	50.70	22.60	11.00	84.30

	Unit	Hours	Hours £	Mat'ls £	O & P £	Total £
one brick thick curved	m2	3.60	60.84	22.60	12.52	95.96
one and a half brick thick	m2	3.70	62.53	33.90	14.46	110.89
two brick thick	m2	4.40	74.36	45.20	17.93	137.49
two brick thick battered	m2	5.00	84.50	45.20	19.46	149.16

Common bricks basic price £200
per thousand in gauged mortar
(1:1:6)

walls

	Unit	Hours	Hours £	Mat'ls £	O & P £	Total £
half brick thick	m2	1.70	28.73	16.45	6.78	51.96
half brick thick overhand	m2	2.20	37.18	16.45	8.04	61.67
half brick thick against						
other work	m2	1.90	32.11	16.45	7.28	55.84
half brick thick curved	m2	2.30	38.87	16.45	8.30	63.62
one brick thick	m2	2.80	47.32	32.90	12.03	92.25
one brick thick curved	m2	3.40	57.46	32.90	13.55	103.91
one and a half brick thick	m2	3.50	59.15	49.35	16.28	124.78
two brick thick	m2	4.20	70.98	65.80	20.52	157.30
two brick thick battered	m2	4.80	81.12	65.80	22.04	168.96

walls, facework one side

	Unit	Hours	Hours £	Mat'ls £	O & P £	Total £
half brick thick	m2	1.80	30.42	16.45	7.03	53.90
half brick thick overhand	m2	2.30	38.87	16.45	8.30	63.62
half brick thick against						
other work	m2	2.00	33.80	16.45	7.54	57.79
half brick thick curved	m2	2.40	40.56	16.45	8.55	65.56
one brick thick	m2	2.90	49.01	32.90	12.29	94.20
one brick thick curved	m2	3.50	59.15	32.90	13.81	105.86
one and a half brick thick	m2	3.60	60.84	49.35	16.53	126.72
two brick thick	m2	4.30	72.67	65.80	20.77	159.24
two brick thick battered	m2	4.90	82.81	65.80	22.29	170.90

walls, facework both sides

	Unit	Hours	Hours £	Mat'ls £	O & P £	Total £
half brick thick	m2	1.90	32.11	16.45	7.28	55.84
half brick thick overhand	m2	2.40	40.56	16.45	8.55	65.56
half brick thick against						
other work	m2	2.10	35.49	16.45	7.79	59.73
half brick thick curved	m2	2.50	42.25	16.45	8.81	67.51

	Unit	Hours	Hours £	Mat'ls £	O & P £	Total £
one brick thick	m2	3.00	50.70	32.90	12.54	96.14
one brick thick curved	m2	3.60	60.84	32.90	14.06	107.80
one and a half brick thick	m2	3.70	62.53	49.35	16.78	128.66
two brick thick	m2	4.40	74.36	65.80	21.02	161.18
two brick thick battered	m2	5.00	84.50	65.80	22.55	172.85

Class B engineering bricks basic
price £250 per thousand in
cement mortar (1:3)

walls						
half brick thick	m2	1.80	30.42	20.45	7.63	58.50
half brick thick overhand	m2	2.30	38.87	20.45	8.90	68.22
half brick thick against						
other work	m2	2.00	33.80	20.45	8.14	62.39
half brick thick curved	m2	2.40	40.56	20.45	9.15	70.16
one brick thick	m2	2.90	49.01	49.90	14.84	113.75
one brick thick curved	m2	3.50	59.15	49.90	16.36	125.41
one and a half brick thick	m2	3.60	60.84	61.35	18.33	140.52
two brick thick	m2	4.30	72.67	81.90	23.19	177.76
two brick thick battered	m2	4.90	82.81	81.90	24.71	189.42

walls, facework one side						
half brick thick	m2	1.90	32.11	20.45	7.88	60.44
half brick thick overhand	m2	2.40	40.56	20.45	9.15	70.16
half brick thick against						
other work	m2	2.10	35.49	20.45	8.39	64.33
half brick thick curved	m2	2.50	42.25	20.45	9.41	72.11
one brick thick	m2	3.00	50.70	49.90	15.09	115.69
one brick thick curved	m2	3.60	60.84	49.90	16.61	127.35
one and a half brick thick	m2	3.70	62.53	61.35	18.58	142.46
two brick thick	m2	4.40	74.36	81.90	23.44	179.70
two brick thick battered	m2	5.00	84.50	81.90	24.96	191.36

walls, facework both sides						
half brick thick	m2	2.00	33.80	20.45	8.14	62.39
half brick thick overhand	m2	2.50	42.25	20.45	9.41	72.11
half brick thick against						
other work	m2	2.20	37.18	20.45	8.64	66.27
half brick thick curved	m2	2.60	43.94	20.45	9.66	74.05

	Unit	Hours	Hours £	Mat'ls £	O & P £	Total £
one brick thick	m2	3.10	52.39	49.90	15.34	117.63
one brick thick curved	m2	3.70	62.53	49.90	16.86	129.29
one and a half brick thick	m2	3.80	64.22	61.35	18.84	144.41
two brick thick	m2	4.50	76.05	81.90	23.69	181.64
two brick thick battered	m2	5.60	94.64	81.90	26.48	203.02

Class B engineering bricks basic
price £350 per thousand in
cement mortar (1:3)

	Unit	Hours	Hours £	Mat'ls £	O & P £	Total £
walls						
half brick thick	m2	1.80	30.42	28.63	8.86	67.91
half brick thick overhand	m2	2.30	38.87	28.63	10.13	77.63
half brick thick against						
other work	m2	2.00	33.80	28.63	9.36	71.79
half brick thick curved	m2	2.40	40.56	28.63	10.38	79.57
one brick thick	m2	2.90	49.01	57.26	15.94	122.21
one brick thick curved	m2	3.50	59.15	57.26	17.46	133.87
one and a half brick thick	m2	3.60	60.84	85.89	22.01	168.74
two brick thick	m2	4.30	72.67	114.57	28.09	215.33
two brick thick battered	m2	4.90	82.81	114.57	29.61	226.99
walls, facework one side						
half brick thick	m2	1.90	32.11	28.63	9.11	69.85
half brick thick overhand	m2	2.40	40.56	28.63	10.38	79.57
half brick thick against						
other work	m2	2.10	35.49	28.63	9.62	73.74
half brick thick curved	m2	2.50	42.25	28.63	10.63	81.51
one brick thick	m2	3.00	50.70	57.26	16.19	124.15
one brick thick curved	m2	3.60	60.84	57.26	17.72	135.82
one and a half brick thick	m2	3.70	62.53	85.89	22.26	170.68
two brick thick	m2	4.40	74.36	114.57	28.34	217.27
two brick thick battered	m2	5.00	84.50	114.57	29.86	228.93
walls, facework both sides						
half brick thick	m2	2.00	33.80	28.63	9.36	71.79
half brick thick overhand	m2	2.50	42.25	28.63	10.63	81.51
half brick thick against						
other work	m2	2.20	37.18	28.63	9.87	75.68
half brick thick curved	m2	2.60	43.94	28.63	10.89	83.46

	Unit	Hours	Hours £	Mat'ls £	O & P £	Total £
one brick thick	m2	3.10	52.39	57.26	16.45	126.10
one brick thick curved	m2	3.70	62.53	57.26	17.97	137.76
one and a half brick thick	m2	3.80	64.22	85.89	22.52	172.63
two brick thick	m2	4.50	76.05	114.57	28.59	219.21
two brick thick battered	m2	5.60	94.64	114.57	31.38	240.59

Facing bricks basic price £250
per 1000 in gauged mortar (1:1:6)

walls

	Unit	Hours	Hours £	Mat'ls £	O & P £	Total £
half brick thick	m2	1.70	28.73	20.30	7.35	56.38
half brick thick overhand	m2	2.20	37.18	20.30	8.62	66.10
half brick thick against other work	m2	1.90	32.11	20.30	7.86	60.27
half brick thick curved	m2	2.30	38.87	20.30	8.88	68.05
one brick thick	m2	2.80	47.32	40.60	13.19	101.11
one brick thick curved	m2	3.40	57.46	40.60	14.71	112.77

Extra over for fair face and flush
pointing

	Unit	Hours	Hours £	Mat'ls £	O & P £	Total £
stretcher bond	m2	0.60	10.14	0.10	1.54	11.78
flemish bond	m2	0.70	11.83	0.10	1.79	13.72
margins	m	0.10	1.69	0.10	0.27	2.06
soffits of flat arches	m	0.65	10.99	0.10	1.66	12.75
soffits of curved arches	m	1.20	20.28	0.10	3.06	23.44

Isolated casings

	Unit	Hours	Hours £	Mat'ls £	O & P £	Total £
half brick thick, facework all round, stretcher bond	m2	2.30	38.87	20.40	8.89	68.16
half brick thick, facework all round, flemish bond	m2	2.40	40.56	20.40	9.14	70.10

Isolated piers

	Unit	Hours	Hours £	Mat'ls £	O & P £	Total £
one brick thick, facework all round, stretcher bond	m2	3.40	57.46	4.08	9.23	70.77
one brick thick, facework all round, flemish bond	m2	3.50	59.15	40.80	14.99	114.94

	Unit	Hours	Hours £	Mat'ls £	O & P £	Total £
Arches						
215mm high 102mm wide	m	1.38	23.32	4.36	4.15	31.83
215mm high 215mm wide	m	2.05	34.65	7.34	6.30	48.28
215mm high, 102mm wide, segmental	m	2.76	46.64	8.27	8.24	63.15
215mm high, 215mm wide, segmental	m	3.72	62.87	11.56	11.16	85.59
Projections						
112mm high, 225mm wide, stretcher bond	m	0.48	8.11	5.14	1.99	15.24
112mm high, 225mm wide, flemish bond	m	0.60	10.14	5.14	2.29	17.57
225mm high, 225mm wide, flemish bond	m	0.85	14.37	10.28	3.70	28.34
112mm high, 328mm wide, stretcher bond	m	0.75	12.68	7.17	2.98	22.82
112mm high, 328mm wide, flemish bond	m	0.95	16.06	7.17	3.48	26.71
Facing bricks basic price £400 per 1000 in gauged mortar (1:1:6)						
walls						
half brick thick	m2	1.70	28.73	28.55	8.59	65.87
half brick thick overhand	m2	2.20	37.18	28.55	9.86	75.59
half brick thick against other work	m2	1.90	32.11	28.55	9.10	69.76
half brick thick curved	m2	2.30	38.87	28.55	10.11	77.53
one brick thick	m2	2.80	47.32	57.10	15.66	120.08
one brick thick curved	m2	3.40	57.46	57.10	17.18	131.74
Extra over for fair face and flush pointing						
stretcher bond	m2	0.60	10.14	0.10	1.54	11.78
flemish bond	m2	0.70	11.83	0.10	1.79	13.72
margins	m	0.10	1.69	0.10	0.27	2.06

	Unit	Hours	Hours £	Mat'ls £	O & P £	Total £
soffits of flat arches	m	0.65	10.99	0.10	1.66	12.75
soffits of curved arches	m	1.20	20.28	0.10	3.06	23.44

Isolated casings

half brick thick, facework all round, stretcher bond	m2	2.30	38.87	28.65	10.13	77.65
half brick thick, facework all round, flemish bond	m2	2.40	40.56	28.65	10.38	79.59

Isolated piers

one brick thick, facework all round, stretcher bond	m2	3.40	57.46	57.30	17.21	131.97
one brick thick, facework all round, flemish bond	m2	3.50	59.15	57.30	17.47	133.92

Arches

215mm high 102mm wide	m	1.38	23.32	6.98	4.55	34.85
215mm high 215mm wide	m	2.05	34.65	11.74	6.96	53.34
215mm high, 102mm wide, segmental	m	2.76	46.64	13.23	8.98	68.86
215mm high, 215mm wide, segmental	m	3.72	0.00 62.87	18.50	12.21	93.57

Projections

112mm high, 225mm wide, stretcher bond	m	0.48	8.11	8.22	2.45	18.78
112mm high, 225mm wide, flemish bond	m	0.60	10.14	8.22	2.75	21.11
225mm high, 225mm wide, flemish bond	m	0.85	14.37	16.38	4.61	35.36
112mm high, 328mm wide, stretcher bond	m	0.75	12.68	11.47	3.62	27.77
112mm high, 328mm wide, flemish bond	m	0.95	16.06	11.47	4.13	31.65

	Unit	Hours	Hours £	Mat'ls £	O & P £	Total £
Facing bricks basic price £500 per 1000 in gauged mortar (1:1:6)						
walls						
half brick thick	m2	1.70	28.73	34.85	9.54	73.12
half brick thick overhand	m2	2.20	37.18	34.85	10.80	82.83
half brick thick against other work	m2	1.90	32.11	34.85	10.04	77.00
half brick thick curved	m2	2.30	38.87	34.85	11.06	84.78
one brick thick	m2	2.80	47.32	69.70	17.55	134.57
one brick thick curved	m2	3.40	57.46	69.70	19.07	146.23
Extra over for fair face and flush pointing						
stretcher bond	m2	0.60	10.14	0.10	1.54	11.78
flemish bond	m2	0.70	11.83	0.10	1.79	13.72
margins	m	0.10	1.69	0.10	0.27	2.06
soffits of flat arches	m	0.65	10.99	0.10	1.66	12.75
soffits of curved arches	m	1.20	20.28	0.10	3.06	23.44
Isolated casings						
half brick thick, facework all round, stretcher bond	m2	2.30	38.87	34.85	11.06	84.78
half brick thick, facework all round, flemish bond	m2	2.40	40.56	34.85	11.31	86.72
Isolated piers						
one brick thick, facework all round, stretcher bond	m2	3.40	57.46	69.70	19.07	146.23
one brick thick, facework all round, flemish bond	m2	3.50	59.15	69.70	19.33	148.18
Arches						
215mm high 102mm wide	m	1.38	23.32	8.73	4.81	36.86
215mm high 215mm wide	m	2.05	34.65	14.68	7.40	56.72

	Unit	Hours	Hours £	Mat'ls £	O & P £	Total £
215mm high, 102mm wide, segmental	m	2.76	46.64	16.54	9.48	72.66
215mm high, 215mm wide, segmental	m	3.72	62.87	23.13	12.90	98.90

Projections

	Unit	Hours	Hours £	Mat'ls £	O & P £	Total £
112mm high, 225mm wide, stretcher bond	m	0.48	8.11	10.28	2.76	21.15
112mm high, 225mm wide, flemish bond	m	0.60	10.14	10.28	3.06	23.48
225mm high, 225mm wide, flemish bond	m	0.85	14.37	16.38	4.61	35.36
112mm high, 328mm wide, stretcher bond	m	0.75	12.68	20.48	4.97	38.13
112mm high, 328mm wide, flemish bond	m	0.95	16.06	14.34	4.56	34.95

Blockwork

Concrete dense aggregate blocks
in gauged mortar (1:1:6)

	Unit	Hours	Hours £	Mat'ls £	O & P £	Total £
walls and partitions						
75mm thick, solid	m2	0.98	16.56	8.38	3.74	28.68
100mm thick, solid	m2	1.14	19.27	7.78	4.06	31.10
140mm thick, solid	m2	1.32	22.31	15.45	5.66	43.42
140mm thick, hollow	m2	1.43	24.17	15.71	5.98	45.86
190mm thick, hollow	m2	1.76	29.74	17.56	7.10	54.40
215mm thick, hollow	m2	2.02	34.14	17.76	7.78	59.68
skins of hollow walls						
75mm thick, solid	m2	1.12	18.93	7.62	3.98	30.53
100mm thick, solid	m2	1.24	20.96	7.78	4.31	33.05
140mm thick, solid	m2	1.27	21.46	15.71	5.58	42.75
215mm thick, hollow	m2	2.15	36.34	17.76	8.11	62.21

	Unit	Hours	Hours £	Mat'ls £	O & P £	Total £
piers and chimney stacks						
75mm thick, solid	m2	1.28	21.63	7.62	4.39	33.64
100mm thick, solid	m2	1.40	23.66	7.78	4.72	36.16
140mm thick, solid	m2	1.68	28.39	15.71	6.62	50.72
isolated casings						
75mm thick, solid	m2	1.42	24.00	7.62	4.74	36.36
100mm thick, solid	m2	1.90	32.11	7.78	5.98	45.87
140mm thick, solid	m2	2.15	36.34	15.71	7.81	59.85

Extra over for fair face and flush pointing

	Unit	Hours	Hours £	Mat'ls £	O & P £	Total £
one side	m2	0.16	2.70	0.00	0.41	3.11
two sides	m2	0.33	5.58	0.00	0.84	6.41

Bonding ends of blockwork to brickwork

	Unit	Hours	Hours £	Mat'ls £	O & P £	Total £
75mm thick, solid	m2	0.30	5.07	0.00	0.76	5.83
100mm thick, solid	m2	0.38	6.42	0.00	0.96	7.39
140mm thick, solid	m2	0.44	7.44	0.00	1.12	8.55
140mm thick, hollow	m2	0.48	8.11	0.00	1.22	9.33
190mm thick, hollow	m2	0.55	9.30	0.00	1.39	10.69
215mm thick, hollow	m2	0.62	10.48	0.00	1.57	12.05

Close cavities at ends of walls and jambs with 100mm thick blockwork and bitumen damp proof course

	Unit	Hours	Hours £	Mat'ls £	O & P £	Total £
bedded in mortar (1:1:6)	m	0.20	3.38	1.56	0.74	5.68

Close cavities at tops of walls with blocks laid flat in mortar (1:1:6)

	Unit	Hours	Hours £	Mat'ls £	O & P £	Total £
100mm thick, solid	m	0.24	4.06	3.12	1.08	8.25
140mm thick, solid	m	0.28	4.73	3.43	1.22	9.39

	Unit	Hours	Hours £	Mat'ls £	O & P £	Total £

Lightweight aerated concrete blocks in mortar (1:1:6)

walls and partitions

	Unit	Hours	Hours £	Mat'ls £	O & P £	Total £
75mm thick	m2	0.90	15.21	7.52	3.41	26.14
100mm thick	m2	1.05	17.75	8.74	3.97	30.46
125mm thick	m2	1.20	20.28	10.05	4.55	34.88
140mm thick	m2	1.26	21.29	14.98	5.44	41.72
150mm thick	m2	1.34	22.65	16.79	5.92	45.35
190mm thick	m2	1.60	27.04	18.32	6.80	52.16
215mm thick	m2	2.00	33.80	21.12	8.24	63.16
255mm thick	m2	2.10	35.49	24.87	9.05	69.41

skins of hollow walls

75mm thick	m2	1.00	16.90	7.52	3.66	28.08
100mm thick	m2	1.15	19.44	8.74	4.23	32.40
125mm thick	m2	1.30	21.97	10.05	4.80	36.82
150mm thick	m2	1.44	24.34	16.29	6.09	46.72

piers and chimney stacks

140mm thick	m2	1.44	24.34	14.98	5.90	45.21
190mm thick	m2	1.52	25.69	20.15	6.88	52.71
215mm thick	m2	1.62	27.38	23.23	7.59	58.20

isolated casings

75mm thick	m2	1.60	27.04	14.98	6.30	48.32
100mm thick	m2	1.92	32.45	20.15	7.89	60.49
140mm thick	m2	2.07	34.98	23.23	8.73	66.94

Extra over for fair face and flush pointing

one side	m2	0.16	2.70	0.00	0.41	3.11
two sides	m2	0.33	5.58	0.00	0.84	6.41

	Unit	Hours	Hours £	Mat'ls £	O & P £	Total £
Bonding ends of blockwork to brickwork						
75mm thick	m	0.25	4.23	0.00	0.63	4.86
100mm thick	m	0.32	5.41	0.00	0.81	6.22
125mm thick	m	0.36	6.08	0.00	0.91	7.00
140mm thick	m	0.40	6.76	0.00	1.01	7.77
150mm thick	m	0.44	7.44	0.00	1.12	8.55
190mm thick	m	0.50	8.45	0.00	1.27	9.72
215mm thick	m	0.55	9.30	0.00	1.39	10.69
255mm thick	m	0.58	9.80	0.00	1.47	11.27
Close cavities at ends of walls and jambs with 100mm thick blockwork and bitumen damp proof course						
bedded in mortar (1:1:6)	m	0.20	3.38	1.69	0.76	5.83
Close cavities at tops of walls with blocks laid flat in mortar (1:1:6)						
100mm thick, solid	m	0.95	16.06	15.77	4.77	36.60
140mm thick, solid	m	1.05	17.75	17.62	5.30	40.67
Lightweight aerated concrete blocks, smooth faced, in mortar (1:1:6)						
walls and partitions						
100mm thick	m2	1.15	19.44	17.40	5.53	42.36
140mm thick	m2	1.36	22.98	24.42	7.11	54.51
190mm thick	m2	1.70	28.73	32.42	9.17	70.32
215mm thick	m2	2.10	35.49	38.72	11.13	85.34
piers and chimney stacks						
140mm thick	m2	1.54	26.03	24.42	7.57	58.01
190mm thick	m2	1.62	27.38	32.42	8.97	68.77
215mm thick	m2	1.72	29.07	37.25	9.95	76.27

	Unit	Hours	Hours £	Mat'ls £	O & P £	Total £
isolated casings						
100mm thick	m2	2.02	34.14	17.40	7.73	59.27
140mm thick	m2	2.16	36.50	24.42	9.14	70.06
Extra over for fair face and flush pointing						
one side	m2	0.16	2.70	0.00	0.41	3.11
two sides	m2	0.33	5.58	0.00	0.84	6.41

Damp proof courses

Hessian based bitumen damp
proof course in gauged mortar
(1:1:6)

	Unit	Hours	Hours £	Mat'ls £	O & P £	Total £
horizontal, width						
over 225mm	m2	0.35	5.92	9.11	2.25	17.28
112mm	m	0.05	0.85	1.06	0.29	2.19
vertical, width						
over 225mm	m2	0.40	6.76	9.11	2.38	18.25
112mm	m	0.07	1.18	1.06	0.34	2.58

Fibre based damp proof course
in gauged mortar (1:1:6)

	Unit	Hours	Hours £	Mat'ls £	O & P £	Total £
horizontal, width						
over 225mm	m2	0.40	6.76	6.36	1.97	15.09
112mm	m	0.05	0.85	0.76	0.24	1.84
vertical, width						
over 225mm	m2	0.45	7.61	6.36	2.09	16.06
112mm	m	0.07	1.18	0.75	0.29	2.22

	Unit	Hours	Hours £	Mat'ls £	O & P £	Total £
Pitch polymer damp proof course in gauged mortar (1:1:6)						
horizontal, width						
over 225mm	m2	0.40	6.76	7.62	2.16	16.54
112mm	m	0.05	0.85	0.90	0.26	2.01
vertical, width						
over 225mm	m2	0.45	7.61	7.62	2.28	17.51
112mm	m	0.07	1.18	0.90	0.31	2.40
Bitumen damp proof course with lead core in mortar (1:1:6)						
horizontal, width						
over 225mm	m2	0.50	8.45	19.32	4.17	31.94
112mm	m	0.06	1.01	2.37	0.51	3.89
vertical, width						
over 225mm	m2	0.55	9.30	19.32	4.29	32.91
112mm	m	0.08	1.35	2.37	0.56	4.28
Two courses of slates bedded in cement mortar (1:3)						
horizontal, width						
over 225mm	m2	0.87	14.70	35.95	7.60	58.25
112mm	m	0.33	5.58	3.47	1.36	10.40
vertical, width						
over 225mm	m2	0.95	16.06	35.95	7.80	59.81
112mm	m	0.35	5.92	3.47	1.41	10.79
Three courses of waterproofing liquid brush on concrete surfaces						
vertically	m2	0.45	7.61	7.97	2.34	17.91
horizontally	m2	0.50	8.45	7.97	2.46	18.88

	Unit	Hours	Hours £	Mat'ls £	O & P £	Total £

Sundries

Expanded polystyrene cavity wall
insulation sheeting, thickness

	Unit	Hours	Hours £	Mat'ls £	O & P £	Total £
25mm	m2	0.20	3.38	5.01	1.26	9.65
50mm	m2	0.22	3.72	7.36	1.66	12.74

Galvanised steel brick
reinforcement, width

	Unit	Hours	Hours £	Mat'ls £	O & P £	Total £
65mm	m	0.07	1.18	0.39	0.24	1.81
115mm	m	0.09	1.52	0.47	0.30	2.29
175mm	m	0.12	2.03	0.76	0.42	3.21
225mm	m	0.15	2.54	1.09	0.54	4.17

Point frames with mastic

	Unit	Hours	Hours £	Mat'ls £	O & P £	Total £
one side	m	0.14	2.37	0.57	0.44	3.38
both sides	m	0.22	3.72	1.14	0.73	5.59

Point frames with polysulphide
sealant

	Unit	Hours	Hours £	Mat'ls £	O & P £	Total £
one side	m	0.14	2.37	1.87	0.64	4.87
both sides	m	0.22	3.72	3.74	1.12	8.58

Rake out joints for flashings and
point up on completion

	Unit	Hours	Hours £	Mat'ls £	O & P £	Total £
horizontal	m	0.09	1.52	0.22	0.26	2.00
stepped	m	0.18	3.04	0.26	0.50	3.80

	Unit	Hours	Hours £	Mat'ls £	O & P £	Total £

MASONRY

Walling

Random rubble walling, laid dry,
thickness

	Unit	Hours	Hours £	Mat'ls £	O & P £	Total £
300mm	m2	3.00	36.00	47.99	12.60	96.59
450mm	m2	3.50	42.00	71.93	17.09	131.02
500mm	m2	3.75	45.00	81.14	18.92	145.06

Random rubble walling, laid dry,
battered one side, thickness

	Unit	Hours	Hours £	Mat'ls £	O & P £	Total £
300mm	m2	3.25	39.00	47.99	13.05	100.04
450mm	m2	3.75	45.00	71.93	17.54	134.47
500mm	m2	4.00	48.00	81.14	19.37	148.51

Random rubble walling, laid dry,
battered both sides, thickness

	Unit	Hours	Hours £	Mat'ls £	O & P £	Total £
300mm	m2	3.50	42.00	47.99	13.50	103.49
450mm	m2	4.00	48.00	71.93	17.99	137.92
500mm	m2	4.25	51.00	81.14	19.82	151.96

Random rubble walling, laid in
gauged mortar (1:1:6), thickness

	Unit	Hours	Hours £	Mat'ls £	O & P £	Total £
300mm	m2	3.20	38.40	51.57	13.50	103.47
450mm	m2	3.70	44.40	77.31	18.26	139.97
500mm	m2	3.90	46.80	86.86	20.05	153.71

Random rubble walling, laid in
gauged mortar (1:1:6), battered
one side, thickness

	Unit	Hours	Hours £	Mat'ls £	O & P £	Total £
300mm	m2	3.45	41.40	51.57	13.95	106.92
450mm	m2	4.00	48.00	77.31	18.80	144.11
500mm	m2	4.15	49.80	86.86	20.50	157.16

	Unit	Hours	Hours £	Mat'ls £	O & P £	Total £
Random rubble walling, laid in gauged mortar (1:1:6), battered both sides, thickness						
300mm	m2	3.95	47.40	51.57	14.85	113.82
450mm	m2	4.50	54.00	77.31	19.70	151.01
500mm	m2	4.65	55.80	86.86	21.40	164.06
Irregular coursed rubble walling, laid in gauged mortar (1:1:6), thickness						
300mm	m2	2.10	25.20	51.57	11.52	88.29
450mm	m2	2.90	34.80	77.31	16.82	128.93
500mm	m2	3.40	40.80	86.86	19.15	146.81
Coursed rubble walling, laid in gauged mortar (1:1:6), thickness						
300mm	m2	2.40	28.80	51.57	12.06	92.43
450mm	m2	3.20	38.40	77.31	17.36	133.07
500mm	m2	3.70	44.40	86.86	19.69	150.95
Fair raking cutting on stone walling, thickness						
300mm	m	0.80	9.60	2.74	1.85	14.19
450mm	m	1.20	14.40	4.44	2.83	21.67
500mm	m	1.80	21.60	5.10	4.01	30.71
Form level bed on stone walling, thickness						
300mm	m	0.20	2.40	0.78	0.48	3.66
450mm	m	0.35	4.20	1.25	0.82	6.27
500mm	m	0.50	6.00	1.37	1.11	8.48

	Unit	Hours	Hours £	Mat'ls £	O & P £	Total £

Sundries

Form holes for pipes up to 50mm
diameter through stone walling,
thickness

300mm	m	1.20	14.40	0.00	2.16	16.56
450mm	m	1.80	21.60	0.00	3.24	24.84
500mm	m	2.65	31.80	0.00	4.77	36.57

Form holes for pipes 50-100mm
diameter through stone walling,
thickness

300mm	m	1.40	16.80	0.00	2.52	19.32
450mm	m	2.00	24.00	0.00	3.60	27.60
500mm	m	2.85	34.20	0.00	5.13	39.33

Form holes for pipes over 100mm
diameter through stone walling,
thickness

300mm	m	2.40	28.80	0.00	4.32	33.12
450mm	m	3.00	36.00	0.00	5.40	41.40
500mm	m	3.85	46.20	0.00	6.93	53.13

Build in ends of steel sections
in stone walling, size

not exceeding 250mm deep	m	0.70	8.40	0.00	1.26	9.66
250-500mm deep	m	0.90	10.80	0.00	1.62	12.42
over 500mm deep	m	1.10	13.20	0.00	1.98	15.18

Mortices in stone walling, size

50 × 50 × 100mm	nr	0.60	7.20	0.00	1.08	8.28
50 × 50 × 150mm	nr	0.65	7.80	0.00	1.17	8.97
75 × 75 × 100mm	nr	0.70	8.40	0.00	1.26	9.66
75 × 75 × 150mm	nr	0.75	7.58	0.00	1.14	8.72

	Unit	Hours	Hours £	Mat'ls £	O & P £	Total £

Grout up mortices in cement
mortar (1:3)

	Unit	Hours	Hours £	Mat'ls £	O & P £	Total £
50 × 50 × 100mm	nr	0.10	1.20	0.11	0.20	1.51
50 × 50 × 150mm	nr	0.15	1.80	0.16	0.29	2.25
75 × 75 × 100mm	nr	0.15	1.80	0.28	0.31	2.39
75 × 75 × 150mm	nr	0.20	2.40	0.33	0.41	3.14

	Unit	Hours	Hours £	Mat'ls £	O & P £	Total £

CARPENTRY AND JOINERY

Sawn softwood, untreated

Floors

50 × 100mm	m	0.16	1.92	1.26	0.48	3.66
50 × 125mm	m	0.17	2.04	1.55	0.54	4.13
50 × 150mm	m	0.18	2.16	1.90	0.61	4.67
75 × 125mm	m	0.17	2.04	2.68	0.71	5.43
75 × 150mm	m	0.18	2.16	3.13	0.79	6.08
75 × 200mm	m	0.19	2.28	3.66	0.89	6.83
75 × 225mm	m	0.20	2.40	4.57	1.05	8.02

Partitions

38 × 75mm	m	0.15	1.80	1.13	0.44	3.37
38 × 100mm	m	0.16	1.92	1.24	0.47	3.63
50 × 75mm	m	0.16	1.92	1.22	0.47	3.61
50 × 100mm	m	0.17	2.04	1.26	0.50	3.80

Flat roofs

38 × 100mm	m	0.10	1.20	1.24	0.37	2.81
50 × 75mm	m	0.11	1.32	1.22	0.38	2.92
50 × 100mm	m	0.12	1.44	1.26	0.41	3.11
50 × 125mm	m	0.13	1.56	1.55	0.47	3.58
50 × 150mm	m	0.15	1.80	1.90	0.56	4.26
75 × 125mm	m	0.16	1.92	2.68	0.69	5.29

Pitched roofs

38 × 100mm	m	0.14	1.68	1.24	0.44	3.36
50 × 75mm	m	0.15	1.80	1.22	0.45	3.47
50 × 100mm	m	0.16	1.92	1.26	0.48	3.66
50 × 125mm	m	0.17	2.04	1.55	0.54	4.13
50 × 150mm	m	0.19	2.28	1.90	0.63	4.81
75 × 125mm	m	0.20	2.40	2.68	0.76	5.84
75 × 150mm	m	0.23	2.77	3.10	0.88	6.75

	Unit	Hours	Hours £	Mat'ls £	O & P £	Total £
Kerbs and bearers						
25 × 75mm	m	0.13	1.56	0.68	0.34	2.58
25 × 100mm	m	0.14	1.68	0.77	0.37	2.82
25 × 150mm	m	0.17	2.04	1.08	0.47	3.59
38 × 75mm	m	0.15	1.80	0.89	0.40	3.09
38 × 100mm	m	0.16	1.92	1.24	0.47	3.63
38 × 150mm	m	0.18	2.16	1.30	0.52	3.98
50 × 50mm	m	0.16	1.92	0.95	0.43	3.30
50 × 75mm	m	0.17	2.04	1.22	0.49	3.75
50 × 100mm	m	0.18	2.16	1.26	0.51	3.93
75 × 75mm	m	0.20	2.40	1.35	0.56	4.31
75 × 100mm	m	0.22	2.64	1.84	0.67	5.15
75 × 125mm	m	0.24	2.88	2.68	0.83	6.39
Solid strutting						
38 × 100mm	m	0.40	4.80	1.24	0.91	6.95
50 × 100mm	m	0.44	5.28	1.26	0.98	7.52
50 × 125mm	m	0.48	5.76	1.42	1.08	8.26
50 × 150mm	m	0.52	6.24	1.55	1.17	8.96
Herringbone strutting, 50 × 50mm to joists, depth						
125mm	m	0.60	7.20	1.33	1.28	9.81
150mm	m	0.60	7.20	1.37	1.29	9.86
175mm	m	0.60	7.20	1.42	1.29	9.91
240mm	m	0.60	7.20	1.46	1.30	9.96
Sawn softwood, impregnated						
Floors						
50 × 100mm	m	0.16	1.92	1.33	0.49	3.74
50 × 125mm	m	0.17	2.04	1.62	0.55	4.21
50 × 150mm	m	0.18	2.16	1.99	0.62	4.77
75 × 125mm	m	0.17	2.04	2.81	0.73	5.58
75 × 150mm	m	0.18	2.16	3.26	0.81	6.23

	Unit	Hours	Hours £	Mat'ls £	O & P £	Total £
75 × 200mm	m	0.19	2.28	3.70	0.90	6.88
75 × 225mm	m	0.20	2.40	4.68	1.06	8.14

Partitions

38 × 75mm	m	0.15	1.80	1.17	0.45	3.42
38 × 100mm	m	0.16	1.92	1.27	0.48	3.67
50 × 75mm	m	0.16	1.92	1.28	0.48	3.68
50 × 100mm	m	0.17	2.04	1.33	0.51	3.88

Flat roofs

38 × 100mm	m	0.10	1.20	1.27	0.37	2.84
50 × 75mm	m	0.11	1.32	1.28	0.39	2.99
50 × 100mm	m	0.12	1.44	1.33	0.42	3.19
50 × 125mm	m	0.13	1.56	1.62	0.48	3.66
50 × 150mm	m	0.15	1.80	1.99	0.57	4.36
75 × 125mm	m	0.16	1.92	2.81	0.71	5.44

Pitched roofs

38 × 100mm	m	0.14	1.68	1.27	0.44	3.39
50 × 75mm	m	0.15	1.80	1.23	0.45	3.48
50 × 100mm	m	0.16	1.92	1.33	0.49	3.74
50 × 125mm	m	0.17	2.04	1.62	0.55	4.21
50 × 150mm	m	0.19	2.28	1.99	0.64	4.91
75 × 125mm	m	0.20	2.40	2.81	0.78	5.99
75 × 150mm	m	0.23	2.77	3.26	0.90	6.94

Kerbs and bearers

25 × 75mm	m	0.13	1.56	0.73	0.34	2.63
25 × 100mm	m	0.14	1.68	0.83	0.38	2.89
25 × 150mm	m	0.17	2.04	1.16	0.48	3.68
38 × 75mm	m	0.15	1.80	0.95	0.41	3.16
38 × 100mm	m	0.16	1.92	1.27	0.48	3.67
38 × 150mm	m	0.18	2.16	1.35	0.53	4.04
50 × 50mm	m	0.16	1.92	0.99	0.44	3.35
50 × 75mm	m	0.17	2.04	1.28	0.50	3.82
50 × 100mm	m	0.18	2.16	1.33	0.52	4.01

	Unit	Hours	Hours £	Mat'ls £	O & P £	Total £
75 × 75mm	m	0.20	2.40	1.40	0.57	4.37
75 × 100mm	m	0.22	2.64	1.90	0.68	5.22
75 × 125mm	m	0.24	2.88	2.81	0.85	6.54

Solid strutting

38 × 100mm	m	0.40	4.80	1.28	0.91	6.99
50 × 100mm	m	0.44	5.28	1.30	0.99	7.57
50 × 125mm	m	0.48	5.76	1.47	1.08	8.31
50 × 150mm	m	0.52	6.24	1.60	1.18	9.02

Herringbone strutting, 50 × 50mm
to joists, depth

125mm	m	0.60	7.20	1.37	1.29	9.86
150mm	m	0.60	7.20	1.43	1.29	9.92
175mm	m	0.60	7.20	1.47	1.30	9.97
240mm	m	0.60	7.20	1.50	1.31	10.01

Gutters and fascias

Marine quality plywood in gutters
over 300mm wide, thickness

12mm	m2	0.90	10.80	12.14	3.44	26.38
18mm	m2	1.00	12.00	16.68	4.30	32.98
25mm	m2	1.10	13.20	19.36	4.88	37.44

Marine quality plywood in gutters
150mm wide, thickness

12mm	m	0.28	3.36	2.02	0.81	6.19
18mm	m	0.32	3.84	2.78	0.99	7.61
25mm	m	0.36	4.32	3.23	1.13	8.68

	Unit	Hours	Hours £	Mat'ls £	O & P £	Total £
Marine quality plywood in gutters 300mm wide, thickness						
12mm	m	0.36	4.32	3.79	1.22	9.33
18mm	m	0.40	4.80	5.50	1.55	11.85
25mm	m	0.44	5.28	6.29	1.74	13.31
Marine quality plywood in eaves, verges, soffits, fascias over 300mm wide, thickness						
12mm	m2	0.80	9.60	12.14	3.26	25.00
18mm	m2	0.90	10.80	16.68	4.12	31.60
25mm	m2	1.00	12.00	19.36	4.70	36.06
Marine quality plywood in gutters 150mm wide, thickness						
12mm	m	0.24	2.88	2.02	0.74	5.64
18mm	m	0.28	3.36	2.78	0.92	7.06
25mm	m	0.32	3.84	3.23	1.06	8.13
Marine quality plywood in gutters 300mm wide, thickness						
12mm	m	0.32	3.33	3.79	1.07	8.19
18mm	m	0.36	3.74	5.50	1.39	10.63
25mm	m	0.40	4.16	6.29	1.57	12.02

Wrought softwood supports

Bearers, framed

	Unit	Hours	Hours £	Mat'ls £	O & P £	Total £
25 × 50mm	m	0.09	0.94	0.42	0.20	1.56
38 × 50mm	m	0.11	1.14	0.66	0.27	2.07
50 × 50mm	m	0.12	1.25	0.77	0.30	2.32
50 × 75mm	m	0.14	1.46	0.86	0.35	2.66
75 × 75mm	m	0.16	1.66	1.07	0.41	3.14

	Unit	Hours	Hours £	Mat'ls £	O & P £	Total £
Bearers, plugged and screwed						
25 × 50mm	m	0.14	1.68	0.53	0.33	2.54
38 × 50mm	m	0.16	1.92	0.77	0.40	3.09
50 × 50mm	m	0.17	2.04	0.88	0.44	3.36
50 × 75mm	m	0.19	2.28	0.97	0.49	3.74
75 × 75mm	m	0.21	2.52	1.18	0.56	4.26

Metal fixings

Galvanised mild steel joist hangers, built in

	Unit	Hours	Hours £	Mat'ls £	O & P £	Total £
38 × 100mm	nr	0.10	1.20	2.27	0.52	3.99
38 × 125mm	nr	0.10	1.20	2.33	0.53	4.06
38 × 150mm	nr	0.10	1.20	2.41	0.54	4.15
38 × 175mm	nr	0.10	1.20	2.44	0.55	4.19
50 × 100mm	nr	0.12	1.44	2.27	0.56	4.27
50 × 125mm	nr	0.12	1.44	2.33	0.57	4.34
50 × 150mm	nr	0.12	1.44	2.41	0.58	4.43
50 × 175mm	nr	0.12	1.44	2.44	0.58	4.46
50 × 200mm	nr	0.12	1.44	2.49	0.59	4.52
50 × 225mm	nr	0.12	1.44	2.55	0.60	4.59
75 × 150mm	nr	0.14	1.68	2.73	0.66	5.07
75 × 175mm	nr	0.14	1.68	2.77	0.67	5.12
75 × 200mm	nr	0.14	1.68	3.04	0.71	5.43
75 × 225mm	nr	0.14	1.68	3.08	0.71	5.47

Galvanised mild steel square toothed timber connectors, single sided, diameter

	Unit	Hours	Hours £	Mat'ls £	O & P £	Total £
38mm	nr	0.03	0.36	0.22	0.09	0.67
50mm	nr	0.03	0.36	0.24	0.09	0.69
63mm	nr	0.03	0.36	0.26	0.09	0.71
75mm	nr	0.03	0.36	0.28	0.10	0.74

	Unit	Hours	Hours £	Mat'ls £	O & P £	Total £
Galvanised mild steel square toothed timber connectors, double sided, diameter						
38mm	nr	0.03	0.36	0.24	0.09	0.69
50mm	nr	0.03	0.36	0.26	0.09	0.71
63mm	nr	0.03	0.36	0.28	0.10	0.74
75mm	nr	0.03	0.36	0.31	0.10	0.77
Galvanised mild steel restraint straps 30 × 2.5mm, girth						
400mm	nr	0.16	1.92	3.17	0.76	5.85
500mm	nr	0.18	2.16	3.44	0.84	6.44
600mm	nr	0.20	2.40	3.94	0.95	7.29
700mm	nr	0.22	2.64	4.29	1.04	7.97
800mm	nr	0.24	2.88	4.60	1.12	8.60
900mm	nr	0.28	3.36	5.17	1.28	9.81
1000mm	nr	0.30	3.60	5.42	1.35	10.37
Galvanised mild steel bolts, 100mm long, diameter						
6mm	nr	0.05	0.60	0.24	0.13	0.97
8mm	nr	0.05	0.60	0.31	0.14	1.05
10mm	nr	0.06	0.72	0.42	0.17	1.31
12mm	nr	0.06	0.72	0.57	0.19	1.48
16mm	nr	0.08	0.96	0.73	0.25	1.94
20mm	nr	0.08	0.96	0.96	0.29	2.21
Galvanised mild steel bolts, 140mm long, diameter						
6mm	nr	0.05	0.60	0.39	0.15	1.14
8mm	nr	0.05	0.60	0.46	0.16	1.22
10mm	nr	0.06	0.72	0.57	0.19	1.48
12mm	nr	0.06	0.72	0.73	0.22	1.67
16mm	nr	0.08	0.96	0.88	0.28	2.12
20mm	nr	0.08	0.96	1.11	0.31	2.38

	Unit	Hours	Hours £	Mat'ls £	O & P £	Total £
Woodwool reinforced slabs 50mm thick, fixed to softwood joists in lengths						
1800mm	m2	0.70	8.40	17.37	3.87	29.64
2100mm	m2	0.70	8.40	17.37	3.87	29.64
2400mm	m2	0.70	8.40	18.11	3.98	30.49
2700mm	m2	0.70	8.40	18.29	4.00	30.69
3000mm	m2	0.70	8.40	18.29	4.00	30.69
Woodwool reinforced slabs 75mm thick, fixed to softwood joists in lengths						
1800mm	m2	0.75	9.00	24.46	5.02	38.48
2100mm	m2	0.75	9.00	24.46	5.02	38.48
2400mm	m2	0.75	9.00	25.41	5.16	39.57
2700mm	m2	0.75	9.00	25.79	5.22	40.01
3000mm	m2	0.75	9.00	25.79	5.22	40.01

Flooring

Wrought softwood boarding, butt-jointed						
19 × 125mm	m2	0.60	7.20	7.27	2.17	16.64
25 × 100mm	m2	0.65	7.80	9.14	2.54	19.48
25 × 150mm	m2	0.60	7.20	9.87	2.56	19.63
Wrought softwood boarding, tongued and grooved						
19 × 125mm	m2	0.70	8.40	7.82	2.43	18.65
25 × 100mm	m2	0.75	9.00	9.17	2.73	20.90
25 × 150mm	m2	0.70	8.40	10.43	2.82	21.65

	Unit	Hours	Hours £	Mat'ls £	O & P £	Total £
Chipboard flooring, butt-jointed, thickness						
18mm	m2	0.35	4.20	3.86	1.21	9.27
25mm	m2	0.40	4.80	4.66	1.42	10.88
Chipboard flooring, tongued and grooved, thickness						
25mm	m2	0.50	6.00	4.95	1.64	12.59

Linings

	Unit	Hours	Hours £	Mat'ls £	O & P £	Total £
Melamine-faced chipboard, 15mm thick to walls						
over 300mm wide	m2	1.10	13.20	4.73	2.69	20.62
less than 300mm wide	m	0.70	8.40	1.78	1.53	11.71
Plain chipboard, 12mm thick to walls						
over 300mm wide	m2	0.48	5.76	2.74	1.28	9.78
less than 300mm wide	m	0.28	3.36	0.86	0.63	4.85
Birch-faced blockboard, 12mm thick to walls						
over 300mm wide	m2	0.60	7.20	10.32	2.63	20.15
less than 300mm wide	m	0.40	4.80	3.44	1.24	9.48
Standard quality hardboard, 3.2mm thick to walls						
over 300mm wide	m2	0.40	4.80	2.10	1.04	7.94
less than 300mm wide	m	0.25	3.00	0.72	0.56	4.28

	Unit	Hours	Hours £	Mat'ls £	O & P £	Total £
Standard quality hardboard, 6mm thick to walls						
over 300mm wide	m2	0.45	5.40	3.29	1.30	9.99
less than 300mm wide	m	0.30	3.60	1.13	0.71	5.44
Internal quality plywood, 4mm thick to walls						
over 300mm wide	m2	0.50	6.00	3.08	1.36	10.44
less than 300mm wide	m	0.30	3.60	1.27	0.73	5.60
Internal quality plywood, 6mm thick to walls						
over 300mm wide	m2	0.55	6.60	3.46	1.51	11.57
less than 300mm wide	m	0.34	4.08	1.42	0.83	6.33
Insulation board, 12mm thick to walls						
over 300mm wide	m2	0.40	4.80	2.13	1.04	7.97
less than 300mm wide	m	0.24	2.88	0.81	0.55	4.24

Windows

Standard softwood windows
without glazing bars, type

	Unit	Hours	Hours £	Mat'ls £	O & P £	Total £
N07V, 488 × 750mm	nr	0.75	9.00	47.29	8.44	64.73
N09V, 488 × 900mm	nr	1.00	12.00	49.63	9.24	70.87
N12V, 488 × 1200mm	nr	1.25	15.00	54.33	10.40	79.73
107C, 630 × 750mm	nr	0.75	9.00	57.53	9.98	76.51
110C, 630 × 1050mm	nr	1.00	12.00	60.13	10.82	82.95
112C, 630 × 1200mm	nr	1.25	15.00	62.16	11.57	88.73
109V, 630 × 900mm	nr	0.75	9.00	60.88	10.48	80.36
110V, 630 × 1050mm	nr	1.25	15.00	63.21	11.73	89.94
112V, 630 × 1200mm	nr	0.75	9.00	64.82	11.07	84.89
2N09W, 915 × 900mm	nr	0.90	10.80	66.04	11.53	88.37

	Unit	Hours	Hours £	Mat'ls £	O & P £	Total £
2N10W, 915 × 1050mm	nr	1.15	13.80	68.57	12.36	94.73
2N12W, 915 × 1200mm	nr	1.40	16.80	71.09	13.18	101.07
212W, 1200 × 1200mm	nr	1.40	16.80	84.73	15.23	116.76
210C, 1200 × 1050mm	nr	1.40	16.80	77.24	14.11	108.15
212T, 1200 × 1200mm	nr	1.40	16.80	94.49	16.69	127.98

Standard hardwood windows
without glazing bars, type

	Unit	Hours	Hours £	Mat'ls £	O & P £	Total £
H2N10W, 915 × 1050mm	nr	1.00	12.00	130.33	21.35	163.68
H2N13W, 915 × 1050mm	nr	1.25	15.00	137.24	22.84	175.08
H2N15W, 915 × 1500mm	nr	1.50	18.00	143.00	24.15	185.15
H213W, 1200 × 1350mm	nr	1.60	19.20	158.22	26.61	204.03
H215W, 1200 × 1500mm	nr	1.70	20.40	163.17	27.54	211.11
H310CC, 1770 × 1050mm	nr	2.00	24.00	240.29	39.64	303.93
H312CC, 1770 × 1200mm	nr	2.20	26.40	260.33	43.01	329.74

PVC-U windows, single glazed,
size

	Unit	Hours	Hours £	Mat'ls £	O & P £	Total £
600 × 900mm	nr	1.00	12.00	54.38	9.96	76.34
600 × 1200mm	nr	1.30	15.60	103.56	17.87	137.03
1200 × 1200mm	nr	1.60	19.20	129.25	22.27	170.72
1800 × 1200mm	nr	2.00	24.00	147.43	25.71	197.14

PVC-U windows, double glazed,
size

	Unit	Hours	Hours £	Mat'ls £	O & P £	Total £
600 × 900mm	nr	1.00	12.00	70.41	12.36	94.77
600 × 1200mm	nr	1.30	15.60	121.78	20.61	157.99
1200 × 1200mm	nr	1.60	19.20	167.95	28.07	215.22
1800 × 1200mm	nr	2.00	24.00	196.35	33.05	253.40

	Unit	Hours	Hours £	Mat'ls £	O & P £	Total £

Doors

Standard internal flush doors,
hardboard faced both sides,
35mm thick, size

	Unit	Hours	Hours £	Mat'ls £	O & P £	Total £
533 × 1981mm	nr	1.20	14.40	26.04	6.07	46.51
610 × 1981mm	nr	1.20	14.40	26.04	6.07	46.51
686 × 1981mm	nr	1.20	14.40	26.04	6.07	46.51
762 × 1981mm	nr	1.20	14.40	26.04	6.07	46.51

Standard internal flush doors,
hardboard faced both sides,
40mm thick, size

	Unit	Hours	Hours £	Mat'ls £	O & P £	Total £
533 × 1981mm	nr	1.20	14.40	28.13	6.38	48.91
610 × 1981mm	nr	1.20	14.40	28.13	6.38	48.91
686 × 1981mm	nr	1.20	14.40	28.13	6.38	48.91
762 × 1981mm	nr	1.20	14.40	28.13	6.38	48.91

Standard internal flush doors,
sapele faced both sides, 35mm
thick, size

	Unit	Hours	Hours £	Mat'ls £	O & P £	Total £
533 × 1981mm	nr	1.30	15.60	39.35	8.24	63.19
610 × 1981mm	nr	1.30	15.60	39.35	8.24	63.19
686 × 1981mm	nr	1.30	15.60	39.35	8.24	63.19
762 × 1981mm	nr	1.30	15.60	39.35	8.24	63.19

Standard internal flush doors,
sapele faced both sides, 40mm
thick, size

	Unit	Hours	Hours £	Mat'ls £	O & P £	Total £
533 × 1981mm	nr	1.30	15.60	43.54	8.87	68.01
610 × 1981mm	nr	1.30	15.60	43.54	8.87	68.01
686 × 1981mm	nr	1.30	15.60	43.54	8.87	68.01
762 × 1981mm	nr	1.30	15.60	43.54	8.87	68.01

	Unit	Hours	Hours £	Mat'ls £	O & P £	Total £

Standard internal flush half-hour
fire check doors, hardboard faced
both sides, 44mm thick, size

686 × 1981mm	nr	1.60	19.20	60.73	11.99	91.92
762 × 1981mm	nr	1.60	19.20	61.95	12.17	93.32
726 × 2040mm	nr	1.60	19.20	64.82	12.60	96.62
826 × 2040mm	nr	1.60	19.20	65.39	12.69	97.28

Standard internal flush half-hour
fire check doors, sapele faced
both sides, 44mm thick, size

686 × 1981mm	nr	1.80	21.60	83.49	15.76	120.85
762 × 1981mm	nr	1.80	21.60	85.47	16.06	123.13
726 × 2040mm	nr	1.80	21.60	88.44	16.51	126.55
826 × 2040mm	nr	1.80	21.60	89.62	16.68	127.90

Standard external flush doors,
plywood faced both sides, 54mm
thick, size

762 × 1981mm	nr	2.10	25.20	58.86	12.61	96.67
838 × 1981mm	nr	2.10	25.20	61.30	12.98	99.48

Framed, ledged and braced doors,
44mm thick

726 × 2040mm	nr	1.30	15.60	77.11	13.91	106.62
826 × 2040mm	nr	1.30	15.60	78.98	14.19	108.77

Door frames and linings

Standard door lining with loose
stops, for door size

686 × 1981mm						
27 × 94mm	nr	0.75	9.00	36.54	6.83	52.37
27 × 107mm	nr	0.75	9.00	38.31	7.10	54.41

	Unit	Hours	Hours £	Mat'ls £	O & P £	Total £
27 × 121mm	nr	0.75	9.00	40.19	7.38	56.57
27 × 133mm	nr	0.75	9.00	42.41	7.71	59.12
762 × 1981mm						
27 × 94mm	nr	0.75	9.00	37.24	6.94	53.18
27 × 107mm	nr	0.75	9.00	39.72	7.31	56.03
27 × 121mm	nr	0.75	9.00	41.55	7.58	58.13
27 × 133mm	nr	0.75	9.00	43.31	7.85	60.16
762 × 2040mm						
27 × 94mm	nr	0.75	9.00	41.92	7.64	58.56
27 × 107mm	nr	0.75	9.00	44.13	7.97	61.10
27 × 121mm	nr	0.75	9.00	47.75	8.51	65.26
27 × 133mm	nr	0.75	9.00	49.17	8.73	66.90

Standard rebated door frame,
for door size

	Unit	Hours	Hours £	Mat'ls £	O & P £	Total £
686 × 1981mm						
35 × 107mm	nr	0.75	9.00	38.31	7.10	54.41
35 × 133mm	nr	0.75	9.00	38.31	7.10	54.41
762 × 1981mm						
35 × 107mm	nr	0.75	9.00	38.31	7.10	54.41
35 × 133mm	nr	0.75	9.00	38.31	7.10	54.41

Wrought softwood rebated
and rounded frames, size

	Unit	Hours	Hours £	Mat'ls £	O & P £	Total £
38 × 100mm	m	0.24	2.88	6.47	1.40	10.75
38 × 125mm	m	0.24	2.88	8.41	1.69	12.98
38 × 140mm	m	0.24	2.88	9.84	1.91	14.63
63 × 88mm	m	0.26	3.12	10.32	2.02	15.46
63 × 100mm	m	0.26	3.12	10.70	2.07	15.89
63 × 125mm	m	0.26	3.12	12.04	2.27	17.43
75 × 100mm	m	0.28	3.36	10.83	2.13	16.32
75 × 125mm	m	0.28	3.36	15.68	2.86	21.90
75 × 140mm	m	0.28	3.36	17.10	3.07	23.53

	Unit	Hours	Hours £	Mat'ls £	O & P £	Total £
Mahogany rebated and rounded frames, size						
38 × 100mm	m	0.30	3.60	13.93	2.63	20.16
38 × 125mm	m	0.30	3.60	15.90	2.93	22.43
38 × 140mm	m	0.30	3.60	18.22	3.27	25.09
63 × 88mm	m	0.32	3.84	19.98	3.57	27.39
63 × 100mm	m	0.32	3.84	20.79	3.69	28.32
63 × 125mm	m	0.32	3.84	23.27	4.07	31.18
75 × 100mm	m	0.36	4.32	22.89	4.08	31.29
75 × 125mm	m	0.36	4.32	30.81	5.27	40.40
75 × 140mm	m	0.36	4.32	37.46	6.27	48.05

Stairs

Wrought softwood open-tread staircase in one flight, 2600mm height, 2700mm going, 12 treads and 13 risers, width

	Unit	Hours	Hours £	Mat'ls £	O & P £	Total £
850 mm	nr	14.00	168.00	383.54	82.73	634.27
910 mm	nr	14.00	168.00	406.30	86.15	660.45

Wrought softwood closed-tread staircase in one flight, 2600mm height, 2700mm going, 12 treads and 13 risers, width

	Unit	Hours	Hours £	Mat'ls £	O & P £	Total £
850 mm	nr	16.00	192.00	544.73	110.51	847.24
910 mm	nr	16.00	192.00	576.58	115.29	883.87

Hardwood open-tread staircase in one flight, 2600mm height, 2700mm going, 12 treads and 13 risers, width

	Unit	Hours	Hours £	Mat'ls £	O & P £	Total £
850 mm	nr	14.00	168.00	710.69	131.80	1010.49
910 mm	nr	14.00	168.00	759.93	139.19	1067.12

	Unit	Hours	Hours £	Mat'ls £	O & P £	Total £

Hardwood closed-tread staircase
in one flight, 2600mm height,
2700mm going, 12 treads and
13 risers, width

	Unit	Hours	Hours £	Mat'ls £	O & P £	Total £
850 mm	nr	16.00	192.00	884.12	161.42	1237.54
910 mm	nr	16.00	192.00	927.66	167.95	1287.61

Kitchen fittings

Fix only kitchen fittings

	Unit	Hours	Hours £	Mat'ls £	O & P £	Total £
wall units 200mm high, size						
300 × 600mm	nr	1.00	12.00	0.00	1.80	13.80
300 × 1000mm	nr	1.10	13.20	0.00	1.98	15.18
300 × 1200mm	nr	1.20	14.40	0.00	2.16	16.56
base units 750mm high, size						
600 × 900mm	nr	1.20	14.40	0.00	2.16	16.56
600 × 1000mm	nr	1.25	15.00	0.00	2.25	17.25
600 × 1200mm	nr	1.30	15.60	0.00	2.34	17.94
900 × 900mm	nr	1.30	15.60	0.00	2.34	17.94
900 × 1000mm	nr	1.35	16.20	0.00	2.43	18.63
900 × 1200mm	nr	1.40	16.80	0.00	2.52	19.32
sink units 750mm high, size						
600 × 900mm	nr	1.30	15.60	0.00	2.34	17.94
600 × 1000mm	nr	1.35	16.20	0.00	2.43	18.63
600 × 1200mm	nr	1.40	16.80	0.00	2.52	19.32
worktops, size						
600 × 900mm	nr	0.40	4.80	0.00	0.72	5.52
600 × 1000mm	nr	0.50	6.00	0.00	0.90	6.90
600 × 1200mm	nr	0.60	7.20	0.00	1.08	8.28

	Unit	Hours	Hours £	Mat'ls £	O & P £	Total £

Insulation

Glass fibre quilt laid over ceiling
joists, thickness

	Unit	Hours	Hours £	Mat'ls £	O & P £	Total £
50mm	m2	0.15	1.80	2.28	0.61	4.69
80mm	m2	0.17	2.04	2.62	0.70	5.36
100mm	m2	0.20	2.40	3.66	0.91	6.97
150mm	m2	0.24	2.88	5.74	1.29	9.91

Glass fibre quilt fixed vertically
to softwood, thickness

	Unit	Hours	Hours £	Mat'ls £	O & P £	Total £
50mm	m2	0.20	2.40	2.28	0.70	5.38
80mm	m2	0.23	2.76	2.62	0.81	6.19

Expanded polystyrene board
fixed vertically to walls with
adhesive, thickness

	Unit	Hours	Hours £	Mat'ls £	O & P £	Total £
12mm	m2	0.50	6.00	1.68	1.15	8.83
25mm	m2	0.60	7.20	3.12	1.55	11.87
50mm	m2	0.70	8.40	5.06	2.02	15.48

Ironmongery

Fix only to softwood

door furniture

	Unit	Hours	Hours £	Mat'ls £	O & P £	Total £
light butts	nr	0.30	3.60	0.00	0.54	4.14
medium butts	nr	0.33	3.96	0.00	0.59	4.55
heavy butts	nr	0.35	4.20	0.00	0.63	4.83
rising butts	nr	0.40	4.80	0.00	0.72	5.52
tee bands	nr	1.00	12.00	0.00	1.80	13.80
barrel bolts, small	nr	0.55	6.60	0.00	0.99	7.59
barrel bolts, medium	nr	0.60	7.20	0.00	1.08	8.28
barrel bolts, large	nr	0.70	8.40	0.00	1.26	9.66
tower bolts, small	nr	0.55	6.60	0.00	0.99	7.59
tower bolts, medium	nr	0.60	7.20	0.00	1.08	8.28
tower bolts, large	nr	0.70	8.40	0.00	1.26	9.66

	Unit	Hours	Hours £	Mat'ls £	O & P £	Total £
panic bolts	nr	2.20	26.40	0.00	3.96	30.36
overhead door closer	nr	1.00	12.00	0.00	1.80	13.80
mortice latch	nr	0.80	9.60	0.00	1.44	11.04
cylinder night rim latch	nr	1.00	12.00	0.00	1.80	13.80
rim dead lock	nr	0.70	8.40	0.00	1.26	9.66
mortice dead lock	nr	0.90	10.80	0.00	1.62	12.42
mortice latch furniture	nr	0.30	3.60	0.00	0.54	4.14
Suffolk latch	nr	0.65	7.80	0.00	1.17	8.97
Norfolk latch	nr	0.65	7.80	0.00	1.17	8.97
postal knocker and plate	nr	1.00	12.00	0.00	1.80	13.80
pull handles	nr	0.25	3.00	0.00	0.45	3.45
push plates	nr	0.25	3.00	0.00	0.45	3.45
window furniture						
casement stay	nr	0.35	4.20	0.00	0.63	4.83
casement fastener	nr	0.25	3.00	0.00	0.45	3.45
cockspur fastener	nr	0.30	3.60	0.00	0.54	4.14
sundries						
numerals	nr	0.05	0.60	0.00	0.09	0.69
hat and coat hook	nr	0.05	0.60	0.00	0.09	0.69
toilet roll holder	nr	0.10	1.20	0.00	0.18	1.38
cabin hook	nr	0.05	0.60	0.00	0.09	0.69
Fix only to hardwood						
door furniture						
light butts	nr	0.40	4.80	0.00	0.72	5.52
medium butts	nr	0.43	5.16	0.00	0.77	5.93
heavy butts	nr	0.45	5.40	0.00	0.81	6.21
rising butts	nr	0.50	6.00	0.00	0.90	6.90
tee bands	nr	1.15	13.80	0.00	2.07	15.87
barrel bolts, small	nr	0.75	9.00	0.00	1.35	10.35
barrel bolts, medium	nr	0.70	8.40	0.00	1.26	9.66
barrel bolts, large	nr	0.80	9.60	0.00	1.44	11.04
tower bolts, small	nr	0.65	7.80	0.00	1.17	8.97
tower bolts, medium	nr	0.70	8.40	0.00	1.26	9.66
tower bolts, large	nr	0.80	9.60	0.00	1.44	11.04
panic bolts	nr	2.40	28.80	0.00	4.32	33.12

	Unit	Hours	Hours £	Mat'ls £	O & P £	Total £
overhead door closer	nr	1.15	13.80	0.00	2.07	15.87
mortice latch	nr	0.90	10.80	0.00	1.62	12.42
cylinder night rim latch	nr	1.15	13.80	0.00	2.07	15.87
rim dead lock	nr	0.80	9.60	0.00	1.44	11.04
mortice dead lock	nr	1.00	12.00	0.00	1.80	13.80
mortice latch furniture	nr	0.40	4.80	0.00	0.72	5.52
Suffolk latch	nr	0.75	9.00	0.00	1.35	10.35
Norfolk latch	nr	0.75	9.00	0.00	1.35	10.35
postal knocker and plate	nr	1.15	13.80	0.00	2.07	15.87
pull handles	nr	0.30	3.60	0.00	0.54	4.14
push plates	nr	0.30	3.60	0.00	0.54	4.14
window furniture						
casement stay	nr	0.45	5.40	0.00	0.81	6.21
casement fastener	nr	0.35	4.20	0.00	0.63	4.83
cockspur fastener	nr	0.40	4.80	0.00	0.72	5.52
sundries						
numerals	nr	0.07	0.84	0.00	0.13	0.97
hat and coat hook	nr	0.07	0.84	0.00	0.13	0.97
toilet roll holder	nr	0.14	1.68	0.00	0.25	1.93
cabin hook	nr	0.07	0.84	0.00	0.13	0.97

	Unit	Hours	Hours £	Mat'ls £	O & P £	Total £

METALWORK

Balustrades

Galvanised steel balustrades

50 × 8mm flat bar	m	0.50	6.00	20.39	3.96	30.35
welded angles	nr	0.75	9.00	0.00	1.35	10.35
ramps	nr	0.75	9.00	0.00	1.35	10.35
flat bends	nr	0.75	9.00	0.00	1.35	10.35
38 × 12mm rails	m	0.50	6.00	20.39	3.96	30.35
welded angles	nr	0.75	9.00	0.00	1.35	10.35
ramps	nr	0.75	9.00	0.00	1.35	10.35
flat bends	nr	0.75	9.00	0.00	1.35	10.35

Lintels

Standard galvanised steel lintels to
100mm thick wall, 143mm high

900mm long	nr	0.15	1.80	6.18	1.20	9.18
1050mm long	nr	0.20	2.40	6.62	1.35	10.37
1200mm long	nr	0.25	3.00	7.27	1.54	11.81

Standard galvanised steel lintels to
256mm thick wall, 143mm high

900mm long	nr	0.20	2.40	34.67	5.56	42.63
1500mm long	nr	0.25	3.00	53.67	8.50	65.17
2100mm long	nr	0.30	3.60	77.16	12.11	92.87

Standard galvanised steel lintels to
256mm thick wall, 219mm high

2700mm long	nr	0.75	9.00	109.29	17.74	136.03
4200mm long	nr	1.00	12.00	259.26	40.69	311.95

	Unit	Hours	Hours £	Mat'ls £	O & P £	Total £

Sundries

Galvanised mild steel water bars, size

	Unit	Hours	Hours £	Mat'ls £	O & P £	Total £
5 × 25mm	m	0.50	6.00	2.97	1.35	10.32
5 × 30mm	m	0.50	6.00	3.57	1.44	11.01
5 × 40mm	m	0.50	6.00	4.49	1.57	12.06

Galvanised steel dowels, size

	Unit	Hours	Hours £	Mat'ls £	O & P £	Total £
8mm diameter × 50mm	nr	0.10	1.20	0.22	0.21	1.63
8mm diameter × 100mm	nr	0.12	1.44	0.28	0.26	1.98
10mm diameter × 50mm	nr	0.10	1.20	0.26	0.22	1.68
10mm diameter × 100mm	nr	0.12	1.44	0.33	0.27	2.04
12mm diameter × 50mm	nr	0.10	1.20	0.28	0.22	1.70
12mm diameter × 100mm	nr	0.12	1.44	0.35	0.27	2.06

	Unit	Hours	Hours £	Mat'ls £	O & P £	Total £

ROOFING

Tiling

Marley Plain granuled or smooth
finish tiles size 267 × 165mm,
65mm lap, 35° pitch, type
1F reinforced underlay

battens size 38 × 19mm

	Unit	Hours	Hours £	Mat'ls £	O & P £	Total £
gauge 100mm	m2	1.35	16.20	26.29	6.37	48.86
gauge 95mm	m2	1.40	16.80	27.50	6.65	50.95
gauge 90mm	m2	1.45	17.40	28.97	6.96	53.33

battens size 38 × 25mm

	Unit	Hours	Hours £	Mat'ls £	O & P £	Total £
gauge 100mm	m2	1.35	16.20	26.63	6.42	49.25
gauge 95mm	m2	1.40	16.80	27.84	6.70	51.34
gauge 90mm	m2	1.45	17.40	29.31	7.01	53.72

Extra for

	Unit	Hours	Hours £	Mat'ls £	O & P £	Total £
nailing every tile with aluminium nails	m2	0.15	1.80	0.44	0.34	2.58
interlocking dry verge	m	0.20	2.40	10.95	2.00	15.35
verge, 150mm wide plain tile undercloak	m	0.20	2.40	2.18	0.69	5.27
double course at eaves	m	0.35	4.20	3.60	1.17	8.97
segmental ridge tile	m	0.40	4.80	9.93	2.21	16.94
valley trough tiles	m	0.60	7.20	14.89	3.31	25.40
segmental hip tiles	m	0.60	7.20	14.89	3.31	25.40
bonnet hip tiles	m	0.60	7.20	14.39	3.24	24.83
eaves vent system	m	0.50	6.00	16.03	3.30	25.33
ventilated ridge terminal	nr	0.50	6.00	37.65	6.55	50.20
gas vent terminal	nr	0.50	6.00	55.51	9.23	70.74
soil vent terminal	nr	0.50	6.00	35.26	6.19	47.45
straight cutting	m	0.20	2.40	0.00	0.36	2.76
holes for pipes	nr	0.40	4.80	0.00	0.72	5.52

	Unit	Hours	Hours £	Mat'ls £	O & P £	Total £
Marley Ludlow Plus smooth finish tiles size 387 × 229mm, battens size 38 × 25mm, type 1F reinforced underlay						
75mm lap, pitch 25-44°	m2	0.90	10.80	11.92	3.41	26.13
100mm lap, pitch 22-44°	m2	0.95	11.40	12.05	3.52	26.97
Extra for						
nailing every tile with aluminium nails	m2	0.10	1.20	0.33	0.23	1.76
interlocking dry verge	m	0.20	2.40	10.95	2.00	15.35
verge, 150mm wide plain tile undercloak	m	0.20	2.40	2.18	0.69	5.27
segmental ridge tile	m	0.40	4.80	9.93	2.21	16.94
valley trough tiles	m	0.60	7.20	14.87	3.31	25.38
segmental hip tiles	m	0.60	7.20	14.87	3.31	25.38
eaves vent system	m	0.50	6.00	16.03	3.30	25.33
ventilated ridge terminal	nr	0.50	6.00	37.65	6.55	50.20
gas vent terminal	nr	0.50	6.00	55.51	9.23	70.74
soil vent terminal	nr	0.50	6.00	35.27	6.19	47.46
straight cutting	m	0.20	2.40	0.00	0.36	2.76
holes for pipes	nr	0.40	4.80	0.00	0.72	5.52
Marley Modern smooth finish tiles size 420 × 330mm, battens size 38 × 25mm, type 1F reinforced underlay						
75mm lap, pitch 22.5-44°	m2	0.80	9.60	12.29	3.28	25.17
Extra for						
nailing every tile with aluminium nails	m2	0.05	0.60	0.22	0.12	0.94
verge, 150mm wide plain tile undercloak	m	0.20	2.40	2.18	0.69	5.27
interlocking dry verge syster	m	0.20	2.40	10.86	1.99	15.25
Modern ridge tile	m	0.60	7.20	10.71	2.69	20.60

	Unit	Hours	Hours £	Mat'ls £	O & P £	Total £
Modern monoridge	m	0.60	7.20	14.42	3.24	24.86
dry ridge system	m	0.50	6.00	12.94	2.84	21.78
Modern hip tiles	m	0.60	7.20	10.06	2.59	19.85
eaves vent system	m	0.50	6.00	16.03	3.30	25.33
ventilated ridge terminal	nr	0.50	6.00	37.65	6.55	50.20
gas vent terminal	nr	0.50	6.00	55.51	9.23	70.74
soil vent terminal	nr	0.50	6.00	35.26	6.19	47.45
straight cutting	m	0.20	2.40	0.00	0.36	2.76
holes for pipes	nr	0.40	4.80	0.00	0.72	5.52

Redland Renown granular faced
or through-coloured tiles size 418
× 330mm, 75mm lap, 343mm
gauge, pitch 30-40°, type
1F reinforced underlay

	Unit	Hours	Hours £	Mat'ls £	O & P £	Total £
battens size 38 × 22mm	m2	0.80	9.60	11.84	3.22	24.66
battens size 38 × 25mm	m2	0.85	10.20	11.87	3.31	25.38

Extra for

	Unit	Hours	Hours £	Mat'ls £	O & P £	Total £
nailing every tile with aluminium nails	m2	0.05	0.60	0.22	0.12	0.94
cloaked verge tile	m	0.20	2.40	9.09	1.72	13.21
half-round ridge or hip tile	m	0.60	7.20	13.16	3.05	23.41
gas flue ridge terminal	nr	0.80	9.60	104.92	17.18	131.70
valley trough tiles	m	0.60	7.20	18.33	3.83	29.36
straight cutting	m	0.20	2.40	0.00	0.36	2.76
holes for pipes	nr	0.40	4.80	0.00	0.72	5.52

Redland Norfolk granular
through-coloured tiles size 381
× 227mm, 100mm lap, 306mm
gauge, pitch 22.5-25°,
type 1F reinforced underlay

	Unit	Hours	Hours £	Mat'ls £	O & P £	Total £
battens size 38 × 22mm	m2	0.85	10.20	15.53	3.86	29.59
battens size 38 × 25mm	m2	0.90	10.80	15.70	3.98	30.48

	Unit	Hours	Hours £	Mat'ls £	O & P £	Total £
Extra for						
nailing every tile with aluminium nails	m2	0.05	0.60	0.22	0.12	0.94
dryvent ridge	m	0.75	9.00	16.66	3.85	29.51
half-round ridge or hip tile	m	0.60	7.20	13.16	3.05	23.41
gas flue ridge terminal	nr	0.80	9.60	102.42	16.80	128.82
valley trough tiles	m	0.60	7.20	18.79	3.90	29.89
straight cutting	m	0.20	2.40	0.00	0.36	2.76
holes for pipes	nr	0.40	4.80	0.00	0.72	5.52

Fibre-cement slating

Asbestos-free artificial slates size
400 × 200mm, pitch 40-45°,
38 × 25mm softwood battens, type
1F reinforced underlay

	Unit	Hours	Hours £	Mat'ls £	O & P £	Total £
lap 70mm, gauge 165mm	m2	1.10	13.20	33.34	6.98	53.52

Asbestos-free artificial slates size
400 × 200mm, pitch 40-45°,
38 × 25mm softwood battens, type
1F reinforced underlay

	Unit	Hours	Hours £	Mat'ls £	O & P £	Total £
lap 90mm, gauge 155mm	m2	1.15	13.80	33..86	2.07	15.87

Asbestos-free artificial slates size
500 × 250mm, pitch over 25°,
38 × 25mm softwood battens, type
1F reinforced underlay

	Unit	Hours	Hours £	Mat'ls £	O & P £	Total £
lap 90mm, gauge 205mm	m2	1.00	12.00	30.81	6.42	49.23

Asbestos-free artificial slates size
600 × 300mm, pitch over 25°,
38 × 25mm softwood battens, type
1F reinforced underlay

	Unit	Hours	Hours £	Mat'ls £	O & P £	Total £
lap 100mm, gauge 250mm	m2	0.90	10.80	26.89	5.65	43.34

	Unit	Hours	Hours £	Mat'ls £	O & P £	Total £
Extra for						
verge, 150mm wide plain tile						
undercloak	m	0.20	2.40	2.10	0.68	5.18
double course at eaves	m	0.35	4.20	6.60	1.62	12.42
half-round ridge tile	m	0.40	4.80	23.78	4.29	32.87
valley tiles	m	0.60	7.20	23.78	4.65	35.63
straight cutting	m	0.20	2.40	0.00	0.36	2.76
holes for pipes	nr	0.40	4.80	0.00	0.72	5.52

Natural slating

Blue/grey Welsh slates size 405 ×
205mm, 75mm lap, 50 × 25mm
softwood battens, type 1F
reinforced underlay

	Unit	Hours	Hours £	Mat'ls £	O & P £	Total £
sloping	m2	1.20	14.40	52.88	10.09	77.37
vertical	m2	1.40	16.80	52.88	10.45	80.13
Extra for						
double eaves course	m	0.50	6.00	24.78	4.62	35.40
single undercloak course	m	0.70	8.40	18.55	4.04	30.99
angled ridge or hip tiles	m	0.70	8.40	16.09	3.67	28.16
mitred hips including cutting both sides	m	0.70	8.40	27.28	5.35	41.03
straight cutting	m	0.60	7.20	0.00	1.08	8.28
holes for pipes	nr	0.40	4.80	0.00	0.72	5.52
fix only lead soakers	nr	0.50	6.00	0.00	0.90	6.90

Blue/grey Welsh slates size 510 ×
255mm, 75mm lap, 50 × 25mm
softwood battens, type 1F
reinforced underlay

	Unit	Hours	Hours £	Mat'ls £	O & P £	Total £
sloping	m2	1.00	12.00	55.11	10.07	77.18
vertical	m2	1.20	14.40	55.11	10.43	79.94

	Unit	Hours	Hours £	Mat'ls £	O & P £	Total £
Extra for						
double eaves course	m	0.50	6.00	31.00	5.55	42.55
single undercloak course	m	0.70	8.40	19.68	4.21	32.29
angled ridge or hip tiles	m	0.70	8.40	16.09	3.67	28.16
mitred hips including cutting						
both sides	m	0.70	8.40	27.28	5.35	41.03
straight cutting	m	0.60	7.20	0.00	1.08	8.28
holes for pipes	nr	0.40	4.80	0.00	0.72	5.52
fix only lead soakers	nr	0.50	6.00	0.00	0.90	6.90

Blue/grey Welsh slates size 610 × 305mm, 75mm lap, 50 × 25mm softwood battens, type 1F reinforced underlay

	Unit	Hours	Hours £	Mat'ls £	O & P £	Total £
sloping	m2	0.80	9.60	58.67	10.24	78.51
vertical	m2	1.00	12.00	58.67	10.60	81.27

Extra for

	Unit	Hours	Hours £	Mat'ls £	O & P £	Total £
double eaves course	m	0.50	6.00	34.98	6.15	47.13
single undercloak course	m	0.70	8.40	21.58	4.50	34.48
angled ridge or hip tiles	m	0.70	8.40	16.09	3.67	28.16
mitred hips including cutting						
both sides	m	0.70	8.40	27.28	5.35	41.03
straight cutting	m	0.60	7.20	0.00	1.08	8.28
holes for pipes	nr	0.40	4.80	0.00	0.72	5.52
fix only lead soakers	nr	0.50	6.00	0.00	0.90	6.90

Westmorland green slates in random lengths 450 - 230mm, 75mm laps, 50 × 25mm softwood battens, type 1F reinforced underlay

	Unit	Hours	Hours £	Mat'ls £	O & P £	Total £
sloping	m2	1.35	16.20	118.21	20.16	154.57
vertical	m2	1.55	18.60	118.21	20.52	157.33

	Unit	Hours	Hours £	Mat'ls £	O & P £	Total £
Extra for						
double eaves course	m	0.50	6.00	34.98	6.15	47.13
single undercloak course	m	0.70	8.40	21.58	4.50	34.48
angled ridge or hip tiles	m	0.70	8.40	16.09	3.67	28.16
straight cutting	m	0.60	7.20	0.00	1.08	8.28
holes for pipes	nr	0.40	4.80	0.00	0.72	5.52
fix only lead soakers	nr	0.50	6.00	0.00	0.90	6.90

Reconstructed stone slating

Marley Monarch interlocking slate
size 325 × 330mm, 38 × 25mm
softwood battens, type 1F
reinforced underlay

	Unit	Hours	Hours £	Mat'ls £	O & P £	Total £
75mm lap, pitch 25-90°	m2	1.00	12.00	31.87	6.58	50.45
100mm lap, pitch 25-90°	m2	1.05	12.60	31.87	6.67	51.14
nailing every tile with aluminium nails	m2	0.05	0.60	0.22	0.12	0.94
interlocking dry verge	m	0.20	2.40	10.95	2.00	15.35
verge, 150mm wide plain tile undercloak	m	0.20	2.40	2.18	0.69	5.27
Modern ridge tiles	m	0.60	7.20	10.06	2.59	19.85
Modern hip tiles	m	0.60	7.20	10.06	2.59	19.85
segmental ridge tile	m	0.40	4.80	9.93	2.21	16.94
Modern monoridge	m	0.60	7.20	14.42	3.24	24.86
dry ridge system	m	0.50	6.00	12.94	2.84	21.78
valley trough tiles	m	0.60	7.20	14.87	3.31	25.38
segmental hip tiles	m	0.60	7.20	14.87	3.31	25.38
ventilated ridge terminal	nr	0.50	6.00	37.65	6.55	50.20
gas vent terminal	nr	0.50	6.00	55.34	9.20	70.54
soil vent terminal	nr	0.50	6.00	35.26	6.19	47.45
straight cutting	m	0.20	2.40	0.00	0.36	2.76
holes for pipes	nr	0.40	4.80	0.00	0.72	5.52

	Unit	Hours	Hours £	Mat'ls £	O & P £	Total £
Redland Cambrian through coloured slates size 300 × 336mm, 38 × 25mm softwood battens, type 1F reinforced underlay						
50mm lap, 250mm gauge	m2	0.80	9.60	29.50	5.87	44.97
90mm lap, 210mm gauge	m2	0.85	10.20	29.98	6.03	46.21
Extra for						
nailing every tile with aluminium nails	m2	0.05	0.60	0.22	0.12	0.94
Extra for						
nailing every tile with aluminium nails	m2	0.05	0.60	0.22	0.12	0.94
dryvent ridge	m	0.75	9.00	16.66	3.85	29.51
half-round ridge or hip tile	m	0.60	7.20	13.16	3.05	23.41
gas flue ridge terminal	nr	0.80	9.60	102.42	16.80	128.82
valley trough tiles	m	0.60	7.20	18.79	3.90	29.89
straight cutting	m	0.20	2.40	0.00	0.36	2.76
holes for pipes	nr	0.40	4.80	0.00	0.72	5.52

Lead sheet coverings

Milled lead roof coverings to flat roofing

	Unit	Hours	Hours £	Mat'ls £	O & P £	Total £
code 5	m2	3.10	37.20	23.27	9.07	69.54
code 6	m2	3.30	39.60	25.70	9.80	75.10

Milled lead roof coverings to dormers

	Unit	Hours	Hours £	Mat'ls £	O & P £	Total £
code 5	m2	3.90	46.80	23.27	10.51	80.58
code 6	m2	4.15	49.80	25.70	11.33	86.83

	Unit	Hours	Hours £	Mat'ls £	O & P £	Total £
Milled lead roof coverings sloping roofing						
code 5	m2	3.30	39.60	23.27	9.43	72.30
code 6	m2	3.50	42.00	25.70	10.16	77.86
Lead flashings, code 5						
horizontal						
150mm girth	m	0.35	4.20	3.50	1.16	8.86
200mm girth	m	0.45	5.40	5.58	1.65	12.63
300mm girth	m	0.60	7.20	7.00	2.13	16.33
sloping						
150mm girth	m	0.55	6.60	3.50	1.52	11.62
200mm girth	m	0.60	7.20	5.58	1.92	14.70
300mm girth	m	0.65	7.80	7.00	2.22	17.02
Lead aprons, code 5						
horizontal						
200mm girth	m	0.65	7.80	5.58	2.01	15.39
300mm girth	m	0.75	9.00	7.00	2.40	18.40
400mm girth	m	0.80	9.60	10.47	3.01	23.08
sloping						
200mm girth	m	0.70	8.40	5.58	2.10	16.08
300mm girth	m	0.75	9.00	7.00	2.40	18.40
400mm girth	m	0.80	9.60	10.47	3.01	23.08
Lead sills, code 5						
horizontal						
200mm girth	m	0.55	6.60	5.58	1.83	14.01
300mm girth	m	0.65	7.80	7.00	2.22	17.02
400mm girth	m	0.75	9.00	10.47	2.92	22.39

	Unit	Hours	Hours £	Mat'ls £	O & P £	Total £
sloping						
200mm girth	m	0.65	7.80	5.58	2.01	15.39
300mm girth	m	0.70	8.40	7.00	2.31	17.71
400mm girth	m	0.80	9.60	10.47	3.01	23.08
Lead hips, code 5						
sloping						
200mm girth	m	0.70	8.40	5.58	2.10	16.08
300mm girth	m	0.90	10.80	7.00	2.67	20.47
400mm girth	m	1.20	14.40	10.47	3.73	28.60
Lead kerbs, code 5						
horizontal						
300mm girth	m	0.65	7.80	7.00	2.22	17.02
400mm girth	m	0.70	8.40	10.47	2.83	21.70
Lead valleys, code 5						
sloping						
400mm girth	m	1.10	13.20	10.47	3.55	27.22
600mm girth	m	1.30	15.60	13.96	4.43	33.99
Lead gutters, code 5						
sloping						
600mm girth	m	1.50	18.00	13.96	4.79	36.75
800mm girth	m	1.80	21.60	18.15	5.96	45.71
Slates, size 400 × 400mm with 200mm high collar, 100mm diameter	nr	1.50	18.00	9.57	4.14	31.71
Slates, size 400 × 400mm with 200mm high collar, 150mm diameter	nr	1.70	20.40	9.64	4.51	34.55

	Unit	Hours	Hours £	Mat'ls £	O & P £	Total £

Copper sheet coverings

Copper sheet 0.55mm thick in
roof coverings to

	Unit	Hours	Hours £	Mat'ls £	O & P £	Total £
flat roof	m2	4.30	51.60	26.24	11.68	89.52
sloping roof	m2	4.50	54.00	26.24	12.04	92.28

Copper sheet 0.61mm thick in
roof coverings to

flat roof	m2	4.40	52.80	27.64	12.07	92.51
sloping roof	m2	4.60	55.20	27.64	12.43	95.27

Copper flashings, 0.55mm thick

horizontal

150mm girth	m	0.45	5.40	3.94	1.40	10.74
200mm girth	m	0.55	6.60	5.25	1.78	13.63
300mm girth	m	0.60	7.20	7.87	2.26	17.33

sloping

150mm girth	m	0.55	6.60	3.94	1.58	12.12
200mm girth	m	0.60	7.20	5.25	1.87	14.32
300mm girth	m	0.65	7.80	7.87	2.35	18.02

Copper flashings, 0.61mm thick

horizontal

150mm girth	m	0.45	5.40	4.14	1.43	10.97
200mm girth	m	0.55	6.60	5.53	1.82	13.95
300mm girth	m	0.60	7.20	8.28	2.32	17.80

sloping

150mm girth	m	0.55	6.60	4.14	1.61	12.35
200mm girth	m	0.60	7.20	5.53	1.91	14.64
300mm girth	m	0.65	7.80	8.28	2.41	18.49

	Unit	Hours	Hours £	Mat'ls £	O & P £	Total £
Copper soakers size						
175 × 175mm	nr	0.25	3.00	1.34	0.65	4.99
300 × 175mm	nr	0.25	3.00	1.68	0.70	5.38
Built-up roofing						
Built-up bituminous felt roof coverings, the layers bonded with hot bitumen						
laid flat						
two layers	m2	0.35	4.20	5.46	1.45	11.11
three layers	m2	0.50	6.00	8.46	2.17	16.63
laid sloping						
two layers	m2	0.40	4.80	5.46	1.54	11.80
three layers	m2	0.55	6.60	8.46	2.26	17.32
Extra for						
aprons to two layer felt						
100mm girth	m	0.25	3.00	1.28	0.64	4.92
150mm girth	m	0.30	3.60	1.84	0.82	6.26
aprons to three layer felt						
100mm girth	m	0.25	3.00	1.51	0.68	5.19
150mm girth	m	0.30	3.60	2.01	0.84	6.45
skirtings to two layer felt						
100mm girth	m	0.25	3.00	1.28	0.64	4.92
150mm girth	m	0.30	3.60	1.84	0.82	6.26
skirtings to three layer felt						
100mm girth	m	0.25	3.00	1.51	0.68	5.19
150mm girth	m	0.30	3.60	2.01	0.84	6.45

	Unit	Hours	Hours £	Mat'ls £	O & P £	Total £
flashings to two layer felt						
100mm girth	m	0.25	3.00	1.08	0.61	4.69
150mm girth	m	0.30	3.60	1.84	0.82	6.26
flashings to three layer felt						
100mm girth	m	0.25	3.00	1.51	0.68	5.19
150mm girth	m	0.30	3.60	2.01	0.84	6.45
Form collars around 100mm diameter pipe						
two layer	nr	1.25	15.00	2.09	2.56	19.65
three layer	nr	1.50	18.00	3.62	3.24	24.86
Form collars around 150mm diameter pipe						
two layer	nr	1.35	16.20	2.25	2.77	21.22
three layer	nr	1.60	19.20	4.08	3.49	26.77

	Unit	Specialist price £	O & P £	Total £

ASPHALT WORK

The following are specialist sub-contractor's prices

Damp proofing and tanking to BS 1097

20mm two coat mastic asphalt coverings to concrete surfaces

flat	m2	30.40	4.56	34.96
sloping 10-45°	m2	33.30	5.00	38.30
sloping 46-90°	m2	36.36	5.45	41.81
vertical	m2	36.36	5.45	41.81

20mm two coat mastic asphalt coverings to concrete surfaces not exceeding 150mm wide

flat	m	4.61	0.69	5.30
sloping 10-45°	m	5.07	0.76	5.83
sloping 46-90°	m	5.62	0.84	6.46
vertical	m	5.62	0.84	6.46

20mm two coat mastic asphalt coverings to concrete surfaces 150-300mm wide

flat	m	9.12	1.37	10.49
sloping 10-45°	m	10.05	1.51	11.56
sloping 46-90°	m	10.97	1.65	12.62
vertical	m	10.97	1.65	12.62

30mm three coat mastic asphalt coverings to concrete surfaces

flat	m2	54.55	8.18	62.73
sloping 10-45 °	m2	57.50	8.63	66.13

	Unit	Specialist price £	O & P £	Total £
sloping 46-90°	m2	61.24	9.19	70.43
vertical	m2	61.24	9.19	70.43

30mm three coat mastic asphalt
coverings to concrete surfaces
not exceeding 150mm wide

	Unit	Specialist price £	O & P £	Total £
flat	m	8.19	1.23	9.42
sloping 10-45°	m	8.96	1.34	10.30
sloping 46-90°	m	9.93	1.49	11.42
vertical	m	9.93	1.49	11.42

30mm three coat mastic asphalt
coverings to concrete surfaces
150-300mm wide

	Unit	Specialist price £	O & P £	Total £
flat	m	16.45	2.47	18.92
sloping 10-45°	m	18.28	2.74	21.02
sloping 46-90°	m	19.93	2.99	22.92
vertical	m	19.93	2.99	22.92

Flooring to BS 1076

20mm two coat mastic asphalt
coverings to flat concrete surfaces

	Unit	Specialist price £	O & P £	Total £
exceeding 300mm	m2	25.02	3.75	28.77
not exceeding 150mm wide	m	3.81	0.57	4.38
150-300mm wide	m	7.69	1.15	8.84

20mm two coat mastic asphalt
coverings to sloping concrete
surfaces

	Unit	Specialist price £	O & P £	Total £
exceeding 300mm	m2	27.76	4.16	31.92
not exceeding 150mm wide	m	4.44	0.67	5.11
150-300mm wide	m	8.75	1.31	10.06

	Unit	Specialist price £	O & P £	Total £
Extra for				
working into recessed frames	m	2.21	0.33	2.54
working into channels 300mm girth	m	4.33	0.65	4.98
working into channels 600mm girth	m	8.82	1.32	10.14
30mm three coat mastic asphalt coverings to flat concrete surfaces				
exceeding 300mm	m2	46.55	6.98	53.53
not exceeding 150mm wide	m	7.07	1.06	8.13
150-300mm wide	m	14.19	2.13	16.32
30mm three coat mastic asphalt coverings to sloping concrete surfaces				
exceeding 300mm	m2	48.77	7.32	56.09
not exceeding 150mm wide	m	7.45	1.12	8.57
150-300mm wide	m	14.67	2.20	16.87
Extra for				
working into recessed frames	m	3.34	0.50	3.84
working into channels 300mm girth	m	7.81	1.17	8.98
working into channels 600mm girth	m	13.70	2.06	15.76

Roofing to BS 988

20mm two coat mastic asphalt coverings

	Unit	Specialist price £	O & P £	Total £
flat	m2	32.41	4.86	37.27
sloping 10-45°	m2	34.22	5.13	39.35
sloping 46-90°	m2	37.59	5.64	43.23
vertical	m2	37.59	5.64	43.23

	Unit	Specialist price £	O & P £	Total £
20mm two coat mastic asphalt coverings to concrete surfaces not exceeding 150mm wide				
flat	m	4.86	0.73	5.59
sloping 10-45°	m	5.12	0.77	5.89
sloping 46-90°	m	5.71	0.86	6.57
vertical	m	5.71	0.86	6.57
20mm two coat mastic asphalt coverings to concrete surfaces 150-300mm wide				
flat	m	9.81	1.47	11.28
sloping 10-45°	m	10.30	1.55	11.85
sloping 46-90°	m	11.31	1.70	13.01
vertical	m	11.31	1.70	13.01

	Unit	Hours	Hours £	Mat'ls £	O & P £	Total £

FLOOR, WALL AND CEILING FINISHINGS

Screeds

Cement and sand (1:3) screed in beds with steel trowelled finish to level floors or to falls not exceeding 15 degrees from the horizontal

	Unit	Hours	Hours £	Mat'ls £	O & P £	Total £
25mm thick						
over 300mm wide	m2	0.30	3.60	2.71	0.95	7.26
not exceeding 300mm wide	m	0.12	1.44	0.92	0.35	2.71
32mm thick						
over 300mm wide	m2	0.32	3.84	3.03	1.03	7.90
not exceeding 300mm wide	m	0.13	1.56	1.09	0.40	3.05
38mm thick						
over 300mm wide	m2	0.33	3.96	3.37	1.10	8.43
not exceeding 300mm wide	m	0.14	1.68	1.63	0.50	3.81
50mm thick						
over 300mm wide	m2	0.36	4.32	4.12	1.27	9.71
not exceeding 300mm wide	m	0.15	1.80	1.43	0.48	3.71
63mm thick						
over 300mm wide	m2	0.40	4.80	5.28	1.51	11.59
not exceeding 300mm wide	m	0.16	1.92	1.85	0.57	4.34
75mm thick						
over 300mm wide	m2	0.75	9.00	6.49	2.32	17.81
not exceeding 300mm wide	m	0.17	2.04	2.08	0.62	4.74

	Unit	Hours	Hours £	Mat'ls £	O & P £	Total £
Cement and sand (1:3) screed in beds with steel trowelled finish to level landings						
25mm thick						
over 300mm wide	m2	0.40	4.80	2.71	1.13	8.64
not exceeding 300mm wide	m	0.15	1.80	0.92	0.41	3.13
32mm thick						
over 300mm wide	m2	0.42	5.04	3.03	1.21	9.28
not exceeding 300mm wide	m	0.16	1.92	1.09	0.45	3.46
38mm thick						
over 300mm wide	m2	0.43	5.16	3.37	1.28	9.81
not exceeding 300mm wide	m	0.17	2.04	1.63	0.55	4.22
50mm thick						
over 300mm wide	m2	0.46	5.52	4.12	1.45	11.09
not exceeding 300mm wide	m	0.18	2.16	1.43	0.54	4.13
63mm thick						
over 300mm wide	m2	0.50	6.00	5.28	1.69	12.97
not exceeding 300mm wide	m	0.19	2.28	1.85	0.62	4.75
75mm thick						
over 300mm wide	m2	0.85	10.20	6.49	2.50	19.19
not exceeding 300mm wide	m	0.33	3.96	2.08	0.91	6.95
Cement and sand (1:3) screed in beds with steel trowelled finish to risers and treads						
25mm thick						
275mm wide	m	0.11	1.32	0.92	0.34	2.58
350mm wide	m	0.13	1.56	1.03	0.39	2.98
32mm thick						
over 300mm wide	m2	0.13	1.56	1.03	0.39	2.98
not exceeding 300mm wide	m	0.15	1.80	1.14	0.44	3.38

	Unit	Hours	Hours £	Mat'ls £	O & P £	Total £
38mm thick						
over 300mm wide	m2	0.16	1.92	1.58	0.53	4.03
not exceeding 300mm wide	m	0.18	2.16	1.68	0.58	4.42
50mm thick						
over 300mm wide	m2	0.17	2.04	1.41	0.52	3.97
not exceeding 300mm wide	m	0.19	2.28	1.52	0.57	4.37

Granolithic, cement and granite
chippings (1:2:5) with steel
trowelled finish to level floors or
to falls not exceeding 15 degrees
from the horizontal

	Unit	Hours	Hours £	Mat'ls £	O & P £	Total £
25mm thick						
over 300mm wide	m2	0.40	4.80	3.92	1.31	10.03
not exceeding 300mm wide	m	0.14	1.68	1.22	0.44	3.34
32mm thick						
over 300mm wide	m2	0.42	5.04	4.25	1.39	10.68
not exceeding 300mm wide	m	0.15	1.80	1.30	0.47	3.57
38mm thick						
over 300mm wide	m2	0.43	5.16	4.59	1.46	11.21
not exceeding 300mm wide	m	0.16	1.92	1.43	0.50	3.85
50mm thick						
over 300mm wide	m2	0.46	5.52	5.31	1.62	12.45
not exceeding 300mm wide	m	0.17	2.04	1.64	0.55	4.23
63mm thick						
over 300mm wide	m2	0.50	6.00	6.54	1.88	14.42
not exceeding 300mm wide	m	0.18	2.16	2.01	0.63	4.80
75mm thick						
over 300mm wide	m2	0.85	10.20	7.85	2.71	20.76
not exceeding 300mm wide	m	0.22	2.64	2.39	0.75	5.78

	Unit	Hours	Hours £	Mat'ls £	O & P £	Total £
Granolithic, cement and granite chippings (1:2:5) with steel trowelled finish to level landings						
25mm thick						
over 300mm wide	m2	0.50	6.00	3.92	1.49	11.41
not exceeding 300mm wide	m	0.16	1.92	1.22	0.47	3.61
32mm thick						
over 300mm wide	m2	0.52	6.24	4.25	1.57	12.06
not exceeding 300mm wide	m	0.17	2.04	1.30	0.50	3.84
38mm thick						
over 300mm wide	m2	0.53	6.36	4.59	1.64	12.59
not exceeding 300mm wide	m	0.18	2.16	1.43	0.54	4.13
50mm thick						
over 300mm wide	m2	0.56	6.72	5.31	1.80	13.83
not exceeding 300mm wide	m	0.19	2.28	1.64	0.59	4.51
Granolithic, cement and granite chippings (1:2:5) with steel trowelled finish to risers and treads						
25mm thick						
275mm wide	m	0.13	1.56	1.25	0.42	3.23
350mm wide	m	0.15	1.80	1.36	0.47	3.63
32mm thick						
over 300mm wide	m2	0.15	1.80	1.36	0.47	3.63
not exceeding 300mm wide	m	0.17	2.04	1.47	0.53	4.04
38mm thick						
over 300mm wide	m2	0.18	2.16	1.96	0.62	4.74
not exceeding 300mm wide	m	0.20	2.40	2.07	0.67	5.14

	Unit	Hours	Hours £	Mat'ls £	O & P £	Total £

In situ wall coverings

Cement and sand (1:3) beds and
backings to brickwork or blockwork walls

13mm thick
over 300mm wide	m2	0.25	3.00	1.14	0.62	4.76
not exceeding 300mm wide	m	0.10	1.20	0.41	0.24	1.85

19mm thick
over 300mm wide	m2	0.28	3.36	2.04	0.81	6.21
not exceeding 300mm wide	m	0.12	1.44	0.73	0.33	2.50

Cement and sand (1:3) beds and
backings to sides of isolated
columns

13mm thick
over 300mm wide	m2	0.35	4.20	1.14	0.80	6.14
not exceeding 300mm wide	m	0.13	1.56	0.41	0.30	2.27

19mm thick
over 300mm wide	m2	0.38	4.56	2.04	0.99	7.59
not exceeding 300mm wide	m	0.15	1.80	0.73	0.38	2.91

Premixed lightweight plaster,
11mm bonding, 2mm finish

brickwork or blockwork walls
over 300mm wide	m2	0.48	5.76	2.22	1.20	9.18
not exceeding 300mm wide	m	0.18	2.16	0.76	0.44	3.36

isolated columns
over 300mm wide	m2	0.58	6.96	2.22	1.38	10.56
not exceeding 300mm wide	m	0.22	2.64	0.76	0.51	3.91

ceilings
over 300mm wide	m2	0.62	7.44	2.04	1.42	10.90
not exceeding 300mm wide	m	0.24	2.88	0.73	0.54	4.15

	Unit	Hours	Hours £	Mat'ls £	O & P £	Total £
One coat 'Universal' plaster to brickwork or blockwork walls						
13mm thick						
over 300mm wide	m2	0.30	3.60	2.84	0.97	7.41
not exceeding 300mm wide	m	0.11	1.32	0.92	0.34	2.58
19mm thick						
over 300mm wide	m2	0.35	4.20	3.52	1.16	8.88
not exceeding 300mm wide	m	0.13	1.56	1.11	0.40	3.07
One coat 'Universal' plaster to isolated columns						
13mm thick						
over 300mm wide	m2	0.40	4.80	2.84	1.15	8.79
not exceeding 300mm wide	m	0.14	1.68	0.92	0.39	2.99
19mm thick						
over 300mm wide	m2	0.45	5.40	3.52	1.34	10.26
not exceeding 300mm wide	m	0.16	1.92	1.11	0.45	3.48
One coat board finish 3mm thick						
plasterboard walls						
over 300mm wide	m2	0.26	3.12	1.32	0.67	5.11
not exceeding 300mm wide	m	0.10	1.20	0.50	0.26	1.96
plasterboard ceilings						
over 300mm wide	m2	0.36	4.32	1.32	0.85	6.49
not exceeding 300mm wide	m	0.14	1.68	0.50	0.33	2.51
plasterboard isolated columns						
over 300mm wide	m2	0.36	4.32	1.32	0.85	6.49
not exceeding 300mm wide	m	0.14	1.68	0.50	0.33	2.51

	Unit	Hours	Hours £	Mat'ls £	O & P £	Total £
Three coat Tyrolean rendering 22mm thick to brickwork or blockwork walls						
over 300mm wide	m2	0.85	10.20	4.59	2.22	17.01
not exceeding 300mm wide	m	0.32	3.84	1.61	0.82	6.27

Lathing

Expamet expanded metal lathing, 6mm thick

	Unit	Hours	Hours £	Mat'ls £	O & P £	Total £
stapled to softwood, vertically						
over 300mm wide	m2	0.64	7.68	5.09	1.92	14.69
not exceeding 300mm wide	m	0.22	2.64	1.52	0.62	4.78
tied with wire, vertically						
over 300mm wide	m2	0.66	7.92	5.09	1.95	14.96
not exceeding 300mm wide	m	0.24	2.88	1.52	0.66	5.06
tied with wire, horizontally						
over 300mm wide	m2	0.74	8.88	5.09	2.10	16.07
not exceeding 300mm wide	m	0.26	3.12	1.52	0.70	5.34

Expamet Riblath expanded metal lathing, 0.3mm thick

	Unit	Hours	Hours £	Mat'ls £	O & P £	Total £
stapled to softwood, vertically						
over 300mm wide	m2	0.54	6.48	6.12	1.89	14.49
not exceeding 300mm wide	m	0.18	2.16	2.00	0.62	4.78
tied with wire, vertically						
over 300mm wide	m2	0.56	6.72	6.12	1.93	14.77
not exceeding 300mm wide	m	0.20	2.40	2.00	0.66	5.06
tied with wire, horizontally						
over 300mm wide	m2	0.64	7.68	6.12	2.07	15.87
not exceeding 300mm wide	m	0.24	2.88	2.00	0.73	5.61

	Unit	Hours	Hours £	Mat'ls £	O & P £	Total £

Plasterboard

Tapered edge wallboard, open
joints to receive skimming or
similar, over 300mm wide

9.5mm thick to

	Unit	Hours	Hours £	Mat'ls £	O & P £	Total £
ceilings over 300mm wide	m2	0.35	4.20	2.13	0.95	7.28
beams not exceeding 600mm	m	0.22	2.64	1.14	0.57	4.35

12.5mm thick to

	Unit	Hours	Hours £	Mat'ls £	O & P £	Total £
ceilings over 300mm wide	m2	0.40	4.80	2.57	1.11	8.48
beams not exceeding 600mm	m	0.25	3.00	1.35	0.65	5.00

Floor tiling

Red clay quarry tiles, bedded in
cement mortar (1:3) 12mm thick,
butt jointed straight both ways

150 × 150 × 12.5mm thick to
floors

	Unit	Hours	Hours £	Mat'ls £	O & P £	Total £
over 300mm wide	m2	0.90	10.80	27.11	5.69	43.60
not exceeding 300mm wide	m	0.36	4.32	8.30	1.89	14.51

200 × 200 × 12.5mm thick to
floors

	Unit	Hours	Hours £	Mat'ls £	O & P £	Total £
over 300mm wide	m2	0.80	9.60	45.04	8.20	62.84
not exceeding 300mm wide	m	0.32	3.84	15.99	2.97	22.80

150 × 150 × 12.5mm thick to
landings

	Unit	Hours	Hours £	Mat'ls £	O & P £	Total £
over 300mm wide	m2	1.00	12.00	27.11	5.87	44.98
not exceeding 300mm wide	m	0.45	5.40	8.30	2.06	15.76

	Unit	Hours	Hours £	Mat'ls £	O & P £	Total £
200 × 200 × 12.5mm thick to landings						
over 300mm wide	m2	0.90	10.80	45.04	8.38	64.22
not exceeding 300mm wide	m	0.36	4.32	15.99	3.05	23.36
Brown clay quarry tiles, bedded in cement mortar (1:3) 12mm thick, butt jointed straight both ways						
150 × 150 × 12.5mm thick to floors						
over 300mm wide	m2	0.90	10.80	32.35	6.47	49.62
not exceeding 300mm wide	m	0.36	4.32	11.19	2.33	17.84
200 × 200 × 12.5mm thick to floors						
over 300mm wide	m2	0.80	9.60	47.76	8.60	65.96
not exceeding 300mm wide	m	0.32	3.84	16.03	2.98	22.85
150 × 150 × 12.5mm thick to landings						
over 300mm wide	m2	1.00	12.00	32.35	6.65	51.00
not exceeding 300mm wide	m	0.49	5.88	11.19	2.56	19.63
200 × 200 × 12.5mm thick to landings						
over 300mm wide	m2	0.90	10.80	47.76	8.78	67.34
not exceeding 300mm wide	m	0.36	4.32	16.03	3.05	23.40
Vinyl floor tiling size 300 × 300mm fixed with adhesive						
2mm thick						
over 300mm wide	m2	0.28	3.36	7.60	1.64	12.60
not exceeding 300mm wide	m	0.10	1.20	2.34	0.53	4.07
2.5mm thick						
over 300mm wide	m2	0.30	3.60	8.74	1.85	14.19
not exceeding 300mm wide	m	0.11	1.32	2.61	0.59	4.52

	Unit	Hours	Hours £	Mat'ls £	O & P £	Total £
3mm thick						
over 300mm wide	m2	0.32	3.84	8.74	1.89	14.47
not exceeding 300mm wide	m	0.12	1.44	2.61	0.61	4.66
Vinyl floor sheeting with welded joints, fixed with adhesive						
2mm thick						
over 300mm wide	m2	0.36	4.32	9.19	2.03	15.54
not exceeding 300mm wide	m	0.14	1.68	3.40	0.76	5.84
2.5mm thick						
over 300mm wide	m2	0.40	4.80	12.50	2.60	19.90
not exceeding 300mm wide	m	0.18	2.16	4.49	1.00	7.65
Linoleum floor sheeting with butt joints, fixed to adhesive						
2.5mm thick						
over 300mm wide	m2	0.34	4.08	9.58	2.05	15.71
not exceeding 300mm wide	m	0.13	1.56	3.85	0.81	6.22
3.2mm thick						
over 300mm wide	m2	0.36	4.32	10.50	2.22	17.04
not exceeding 300mm wide	m	0.14	1.68	3.89	0.84	6.41

Wall tiling

White glazed ceramic wall tiling, fixed with adhesive, pointing with white grout

	Unit	Hours	Hours £	Mat'ls £	O & P £	Total £
108 × 108 × 4mm thick						
over 300mm wide	m2	0.95	11.40	19.48	4.63	35.51
not exceeding 300mm wide	m	0.38	4.56	6.75	1.70	13.01

	Unit	Hours	Hours £	Mat'ls £	O & P £	Total £

152 × 152 × 5.6mm thick

	Unit	Hours	Hours £	Mat'ls £	O & P £	Total £
over 300mm wide	m2	0.85	10.20	18.24	4.27	32.71
not exceeding 300mm wide	m	0.34	4.08	6.27	1.55	11.90

203 × 102 × 6.5mm thick

	Unit	Hours	Hours £	Mat'ls £	O & P £	Total £
over 300mm wide	m2	0.75	9.00	17.11	3.92	30.03
not exceeding 300mm wide	m	0.30	3.60	6.04	1.45	11.09

Extra for

	Unit	Hours	Hours £	Mat'ls £	O & P £	Total £
cutting to edges	m	0.10	1.20	0.00	0.18	1.38
holes for small pipes	nr	0.20	2.40	0.00	0.36	2.76
holes for large pipes	nr	0.30	3.60	0.00	0.54	4.14

Dark-coloured glazed ceramic wall
tiling, fixed with adhesive, pointing
with matching grout

108 × 108 × 4mm thick

	Unit	Hours	Hours £	Mat'ls £	O & P £	Total £
over 300mm wide	m2	0.95	11.40	24.35	5.36	41.11
not exceeding 300mm wide	m	0.38	4.56	8.21	1.92	14.69

152 × 152 × 5.6mm thick

	Unit	Hours	Hours £	Mat'ls £	O & P £	Total £
over 300mm wide	m2	0.85	10.20	23.14	5.00	38.34
not exceeding 300mm wide	m	0.34	4.08	7.63	1.76	13.47

203 × 102 × 6.5mm thick

	Unit	Hours	Hours £	Mat'ls £	O & P £	Total £
over 300mm wide	m2	0.75	9.00	21.91	4.64	35.55
not exceeding 300mm wide	m	0.30	3.60	7.13	1.61	12.34

Extra for

	Unit	Hours	Hours £	Mat'ls £	O & P £	Total £
cutting to edges	m	0.10	1.20	0.00	0.18	1.38
holes for small pipes	nr	0.20	2.40	0.00	0.36	2.76
holes for large pipes	nr	0.30	3.60	0.00	0.54	4.14

	Unit	Hours	Hours £	Mat'ls £	O & P £	Total £

PLUMBING AND HEATING

Sanitary fittings

Acrylic reinforced bath, size 1700
× 700mm, complete with 2nr
chromium plated grips, 40mm
waste fitting, overflow with chain
and plastic plug

	Unit	Hours	Hours £	Mat'ls £	O & P £	Total £
white	nr	3.00	43.50	147.94	28.72	220.16
coloured	nr	3.00	43.50	161.28	30.72	235.50

Porcelain enamel standard gauge
bath, size 1700 × 700mm, complete
with 2nr chromium plated grips,
40mm waste fitting, overflow
with chain and plastic plug

	Unit	Hours	Hours £	Mat'ls £	O & P £	Total £
white	nr	2.90	42.05	141.23	27.49	210.77
coloured	nr	2.90	42.05	153.81	29.38	225.24

Bath panels, moulded acrylic,
fixing in position for trimming as
as required

	Unit	Hours	Hours £	Mat'ls £	O & P £	Total £
end panel	nr	0.30	4.35	12.55	2.54	19.44
side panel	nr	0.50	7.25	14.99	3.34	25.58

Angle strip polished aluminium,
fixing with chromium plated dome
headed screws, cutting to length

	Unit	Hours	Hours £	Mat'ls £	O & P £	Total £
25 × 25 × 560mm long	nr	0.35	5.08	4.55	1.44	11.07

Wash basin, vitreous china, size
510 × 410mm, complete with
pedestal, chromium plated waste,
overflow with chain and plastic
plug

	Unit	Hours	Hours £	Mat'ls £	O & P £	Total £
white	nr	2.20	31.90	82.46	17.15	131.51
coloured	nr	2.20	31.90	89.75	18.25	139.90

	Unit	Hours	Hours £	Mat'ls £	O & P £	Total £
Wash basin, vitreous china, size 610 × 470mm, complete with brackets, chromium plated waste, overflow with chain and plastic plug						
white	nr	2.30	33.35	85.61	17.84	136.80
coloured	nr	2.30	33.35	92.55	18.89	144.79
Stainless steel single roll-edge type sink, complete with inset chromium plated waste, overflow with chain and plastic plug, size						
1000 × 500mm	nr	1.30	18.85	89.05	16.19	124.09
1000 × 600mm	nr	1.40	20.30	96.61	17.54	134.45
1200 × 600mm	nr	1.50	21.75	108.04	19.47	149.26
Stainless steel double sit-on type sink, complete with inset chromium plated waste, overflow with chain and plastic plug, size						
1500 × 500mm	nr	1.80	26.10	107.54	20.05	153.69
1500 × 600mm	nr	1.90	27.55	123.64	22.68	173.87
Vitreous china low level WC suite comprising pan, plastic seat and cover, 9 litre cistern, low pressure ball valve, connecting pipework screwed to floor	nr	2.35	34.08	184.67	32.81	251.56
Vitreous china low level close coupled WC suite comprising pan, plastic seat and cover, 9 litre cistern, low pressure ball valve, connecting pipework screwed to floor	nr	2.35	34.08	204.56	35.80	274.43
Free-standing plain rim, vitreous china bidet excluding fittings						
white	nr	2.50	36.25	258.79	44.26	339.30
coloured	nr	2.50	36.25	278.40	47.20	361.85

	Unit	Hours	Hours £	Mat'ls £	O & P £	Total £
Bowl urinals, white vitreous china, wall-mounted on hangars chromium plated dome outlet						
430mm wide × 305mm high	nr	1.00	14.50	110.14	18.70	143.34
500mm wide × 350mm high	nr	1.10	15.95	152.67	25.29	193.91
Chromium plated pillar taps						
13mm	pr	0.40	5.80	23.88	4.45	34.13
19mm	pr	0.45	6.53	31.92	5.77	44.21
Chromium plated mixer taps, 19mm						
cross-top handles	pr	0.60	8.70	58.93	10.14	77.77
lever handles	pr	0.60	8.70	79.38	13.21	101.29
Chromium plated thermostatic exposed shower valve with flexible hose, slide rail and spray set	nr	1.40	20.30	201.56	33.28	255.14
Instant electric 9.5kw shower with flexible hose, slide rail and spray set	nr	1.50	21.75	167.28	28.35	217.38
Shower cubicle size 788 × 842 × 2115mm in anodised aluminium frame and safety glass	nr	2.00	29.00	689.48	107.77	826.25
White acrylic fireclay shower tray size 750 × 750 × 180mm with chromium plated waste fitting	nr	1.30	18.85	98.35	17.58	134.78

Holes and chases

	Unit	Hours	Hours £	Mat'ls £	O & P £	Total £
Cutting holes for pipes up to 25mm diameter in walls 102.5mm thick						
commons	nr	0.40	5.80	0.00	0.87	6.67
facings	nr	0.70	10.15	0.00	1.52	11.67
engineering class A	nr	0.90	13.05	0.00	1.96	15.01

	Unit	Hours	Hours £	Mat'ls £	O & P £	Total £
Cutting holes for pipes up to 25mm diameter in walls 100mm thick						
concrete blocks	nr	0.30	4.35	0.00	0.65	5.00
Thermalite blocks	nr	0.20	2.90	0.00	0.44	3.34
aerated concrete blocks	nr	0.25	3.63	0.00	0.54	4.17
Cutting and pinning ends of pipe support brackets and make good to						
concrete	nr	0.45	6.53	0.00	0.98	7.50
brickwork	nr	0.35	5.08	0.00	0.76	5.84
blockwork	nr	0.30	4.35	0.00	0.65	5.00
tiled walls	nr	0.50	7.25	0.00	1.09	8.34
Cut opening through cavity wall comprising facing bricks and concrete blocks						
balanced flue outlet	nr	0.90	13.05	0.00	1.96	15.01
150mm diameter flue pipe	nr	0.75	10.88	0.00	1.63	12.51
19mm overflow pipe	nr	0.20	2.90	0.00	0.44	3.34
Cutting chases for one pipe up to 25mm diameter in						
commons	nr	0.35	5.08	0.00	0.76	5.84
facings	nr	0.38	5.51	0.00	0.83	6.34
engineering class A	nr	0.45	6.53	0.00	0.98	7.50
engineering class B	nr	0.40	5.80	0.00	0.87	6.67
concrete blocks	nr	0.35	5.08	0.00	0.76	5.84
Thermalite blocks	nr	0.25	3.63	0.00	0.54	4.17
aerated concrete blocks	nr	0.20	2.90	0.00	0.44	3.34
plastered surfaces	m	0.15	2.18	0.00	0.33	2.50
tiled surfaces	m	0.25	3.63	0.00	0.54	4.17
concrete floor	m	0.35	5.08	0.00	0.76	5.84
granolithic floor	m	0.50	7.25	0.00	1.09	8.34
concrete soffit	m	0.55	7.98	0.00	1.20	9.17

	Unit	Hours	Hours £	Mat'ls £	O & P £	Total £

Rainwater pipes and gutters

PVC-U rainwater pipe plugged
to brickwork with pipe brackets
and fitting clips at 2m maximum
centres

	Unit	Hours	Hours £	Mat'ls £	O & P £	Total £
68mm diameter pipe	m	0.25	3.63	5.61	1.39	10.62
extra over for						
bend, 87.5°	nr	0.25	3.63	5.39	1.35	10.37
offset	nr	0.25	3.63	3.21	1.03	7.86
branch	nr	0.25	3.63	10.73	2.15	16.51
shoe	nr	0.25	3.63	4.65	1.24	9.52
access pipe	nr	0.25	3.63	14.77	2.76	21.15
hopper head	nr	0.25	3.63	17.26	3.13	24.02
68mm square pipe	m	0.25	3.63	6.72	1.55	11.90
extra over for						
bend, 87.5°	nr	0.25	3.63	3.44	1.06	8.12
offset	nr	0.25	3.63	6.04	1.45	11.11
branch	nr	0.25	3.63	12.12	2.36	18.11
shoe	nr	0.25	3.63	3.95	1.14	8.71
access pipe	nr	0.25	3.63	18.44	3.31	25.37
hopper head	nr	0.25	3.63	17.26	3.13	24.02

Cast iron rainwater pipe,
straight, ears cast on, spigot
and socket dry joints plugged
to brickwork

	Unit	Hours	Hours £	Mat'ls £	O & P £	Total £
75mm diameter pipe	m	0.30	4.35	20.51	3.73	28.59
extra over for						
offset, 150mm	nr	0.30	4.35	15.27	2.94	22.56
offset, 229mm	nr	0.30	4.35	17.79	3.32	25.46
offset, 305mm	nr	0.30	4.35	21.86	3.93	30.14
bend	nr	0.30	4.35	12.12	2.47	18.94
branch	nr	0.30	4.35	21.22	3.84	29.41
eared shoe	nr	0.30	4.35	16.30	3.10	23.75
flat hopper	nr	0.30	4.35	14.47	2.82	21.64
rectangular hopper	nr	0.30	4.35	56.74	9.16	70.25

	Unit	Hours	Hours £	Mat'ls £	O & P £	Total £
100mm diameter pipe	m	0.35	5.08	26.80	4.78	36.66
extra over for						
offset, 150mm	nr	0.35	5.08	28.83	5.09	38.99
offset, 225mm	nr	0.35	5.08	34.91	6.00	45.98
offset, 305mm	nr	0.35	5.08	56.74	9.27	71.09
bend	nr	0.35	5.08	17.12	3.33	25.52
branch	nr	0.35	5.08	25.21	4.54	34.83
eared shoe	nr	0.35	5.08	21.66	4.01	30.75
flat hopper	nr	0.35	5.08	32.06	5.57	42.71
rectangular hopper	nr	0.35	5.08	56.74	9.27	71.09
PVC-U half round rainwater gutter, fixed to timber with support brackets at 1m maximum centres						
76mm gutter	m	0.20	2.90	6.07	1.35	10.32
extra over for						
running outlet	nr	0.20	2.90	5.53	1.26	9.69
angle	nr	0.20	2.90	6.03	1.34	10.27
stop end outlet	nr	0.10	1.45	4.85	0.95	7.25
stop end	nr	0.10	1.45	2.97	0.66	5.08
112mm gutter	m	0.26	3.77	4.48	1.24	9.49
extra over for						
running outlet	nr	0.26	3.77	3.90	1.15	8.82
angle	nr	0.26	3.77	4.45	1.23	9.45
stop end outlet	nr	0.13	1.89	3.90	0.87	6.65
stop end	nr	0.13	1.89	1.95	0.58	4.41
Cast iron half round rainwater gutter with mastic joints, primed, fixed to timber with support brackets at 1m maximum centres						
100mm gutter	m	0.36	5.22	10.85	2.41	18.48
extra over for						
running outlet	nr	0.36	5.22	6.97	1.83	14.02
angle	nr	0.36	5.22	7.16	1.86	14.24

	Unit	Hours	Hours £	Mat'ls £	O & P £	Total £
Plastic wire balloon guard for pipes and outlets, diameter						
50mm	nr	0.05	0.73	1.97	0.40	3.10
63mm	nr	0.05	0.73	2.00	0.41	3.13
75mm	nr	0.05	0.73	2.15	0.43	3.31
100mm	nr	0.05	0.73	3.02	0.56	4.31

Waste pipes

MuPVC waste system, solvent welded joints, clips at 500mm maximum centres, plugged to brickwork

	Unit	Hours	Hours £	Mat'ls £	O & P £	Total £
32mm diameter pipe	m	0.25	3.63	3.02	1.00	7.64
extra over for						
bend, 45°	nr	0.24	3.48	1.84	0.80	6.12
bend, 87.5°	nr	0.24	3.48	1.84	0.80	6.12
long tail bend, 90°	nr	0.24	3.48	1.84	0.80	6.12
tee	nr	0.24	3.48	3.14	0.99	7.61
bottle trap	nr	0.24	3.48	4.27	1.16	8.91
tubular P trap	nr	0.24	3.48	4.20	1.15	8.83
tubular S trap	nr	0.24	3.48	5.03	1.28	9.79
50mm diameter pipe	m	0.34	4.93	4.57	1.43	10.93
extra over for						
bend, 45°	nr	0.32	4.64	2.95	1.14	8.73
bend, 87.5°	nr	0.32	4.64	2.95	1.14	8.73
long tail bend, 90°	nr	0.32	4.64	2.95	1.14	8.73
tee	nr	0.32	4.64	6.34	1.65	12.63
cross tee	nr	0.32	4.64	11.74	2.46	18.84

	Unit	Hours	Hours £	Mat'ls £	O & P £	Total £

MuPVC waste system, push-fit joints, clips at 500mm maximum centres, plugged to brickwork

	Unit	Hours	Hours £	Mat'ls £	O & P £	Total £
32mm diameter pipe	m	0.20	2.90	1.09	0.60	4.59
extra over for						
bend, 45°	nr	0.18	2.61	0.92	0.53	4.06
knuckle bend, 90°	nr	0.18	2.61	0.90	0.53	4.04
spigot bend, 90°	nr	0.18	2.61	1.13	0.56	4.30
tee	nr	0.18	2.61	1.44	0.61	4.66
P trap, 38mm	nr	0.18	2.61	4.20	1.02	7.83
bottle trap, 38mm	nr	0.18	2.61	4.27	1.03	7.91
bottle trap, 76mm	nr	0.18	2.61	6.82	1.41	10.84
anti-syphon trap, 76mm	nr	0.18	2.61	4.99	1.14	8.74
connection to back inlet						
gulley, caulking bush	nr	0.16	2.32	2.87	0.78	5.97
40mm diameter pipe	m	0.24	3.48	1.84	0.80	6.12
extra over for						
bend, 45°	nr	0.20	2.90	0.98	0.58	4.46
knuckle bend, 90°	nr	0.20	2.90	0.92	0.57	4.39
spigot bend, 90°	nr	0.20	2.90	1.19	0.61	4.70
tee	nr	0.20	2.90	1.48	0.66	5.04

Soil pipes

PVC-U soil system, solvent-welded joints, holderbats at 1250mm maximum centres, plugged to brickwork

	Unit	Hours	Hours £	Mat'ls £	O & P £	Total £
110mm diameter pipe	m	0.38	5.51	11.53	2.56	19.60
extra over for						
bend, 92.5°	nr	0.34	4.93	14.82	2.96	22.71
bend, 135°	nr	0.34	4.93	14.82	2.96	22.71
bend, variable	nr	0.34	4.93	14.82	2.96	22.71
bend, spigot/spigot	nr	0.34	4.93	14.82	2.96	22.71

	Unit	Hours	Hours £	Mat'ls £	O & P £	Total £
110mm diameter pipe (cont'd)						
branch, single, 92.5°	nr	0.40	5.80	19.31	3.77	28.88
branch, single, 135°	nr	0.40	5.80	19.31	3.77	28.88
branch, single, 92.5° spigot outlet	nr	0.40	5.80	19.31	3.77	28.88
branch, single, 92.5° socket outlet	nr	0.40	5.80	19.31	3.77	28.88
access pipe connector	nr	0.34	4.93	19.80	3.71	28.44
access pipe, single branch	nr	0.34	4.93	19.31	3.64	27.88
access cap	nr	0.38	5.51	11.89	2.61	20.01
WC manifold connector	nr	0.38	5.51	9.07	2.19	16.77
vent cowl	nr	0.22	3.19	3.56	1.01	7.76
Cast iron soil system with flexible joints, pipe brackets at 2m maximum centres, plugged to brickwork						
100mm diameter pipe	m	0.65	9.43	29.47	5.83	44.73
extra over for						
bend, short radius	nr	0.55	7.98	13.62	3.24	24.83
bend, short radius with door	nr	0.55	7.98	28.80	5.52	42.29
bend, long radius	nr	0.55	7.98	22.06	4.51	34.54
bend, long radius, access	nr	0.55	7.98	37.26	6.79	52.02
bend, long radius, heel rest	nr	0.55	7.98	26.71	5.20	39.89
bend, long tail, 87.5 °	nr	0.55	7.98	17.60	3.84	29.41
branch, single, plain	nr	0.60	8.70	21.05	4.46	34.21
branch, single with door	nr	0.60	8.70	36.27	6.75	51.72
branch, double, plain	nr	0.60	8.70	26.05	5.21	39.96
branch, double with door	nr	0.60	8.70	41.24	7.49	57.43
branch, corner	nr	0.60	8.70	34.27	6.45	49.42
offset, 75mm projection	nr	0.55	6.02	14.33	3.05	23.40
offset, 115mm projection	nr	0.55	7.98	17.08	3.76	28.81
offset, 150mm projection	nr	0.55	7.98	17.08	3.76	28.81
offset, 225mm projection	nr	0.55	7.98	19.58	4.13	31.69

	Unit	Hours	Hours £	Mat'ls £	O & P £	Total £

100mm diameter pipe (cont'd)

	Unit	Hours	Hours £	Mat'ls £	O & P £	Total £
offset, 300mm projection	nr	0.55	7.98	22.06	4.51	34.54
blank end, plain	nr	0.55	7.98	4.15	1.82	13.94
blank end, drilled and tapped	nr	0.55	7.98	19.54	4.13	31.64
P trap, plain	nr	0.55	7.98	4.28	1.84	14.09
P trap with door	nr	0.55	7.98	51.76	8.96	68.70
WC connector, 305mm effective length	nr	0.55	7.98	23.47	4.72	36.16

Overflows

MPVC-U overflow system, solvent-welded joints, clips at 500mm maximum centres, plugged to brickwork

	Unit	Hours	Hours £	Mat'ls £	O & P £	Total £
19mm diameter pipe	m	0.20	2.90	1.72	0.69	5.31
extra over for						
bend	nr	0.18	2.61	1.26	0.58	4.45
tee	nr	0.18	2.61	1.31	0.59	4.51
connector, bent tank	nr	0.20	2.90	1.87	0.72	5.49

Traps

Polypropylene traps, screwed joints to outlet and pipe

	Unit	Hours	Hours £	Mat'ls £	O & P £	Total £
Bottle trap, 38mm seal						
32mm	nr	0.30	4.35	4.27	1.29	9.91
40mm	nr	0.35	5.08	4.35	1.41	10.84
Anti-syphon bottle trap						
32mm	nr	0.30	4.35	6.82	1.68	12.85
40mm	nr	0.35	5.08	7.09	1.82	13.99
Tubular S trap						
32mm	nr	0.30	4.35	5.03	1.41	10.79
40mm	nr	0.35	5.08	5.99	1.66	12.72

	Unit	Hours	Hours £	Mat'ls £	O & P £	Total £
Tubular P trap						
32mm	nr	0.30	4.35	6.82	1.68	12.85
40mm	nr	0.35	5.08	7.09	1.82	13.99
Running P trap						
40mm	nr	0.35	5.08	7.20	1.84	14.12
Bath trap						
40mm	nr	0.35	5.08	6.32	1.71	13.10
Bath trap with overflow						
40mm	nr	0.35	5.08	5.90	1.65	12.62

Copper pipework, capillary joints

Copper pipe, lead-free
pre-soldered capillary joints
and fittings, clips at 1250mm
centres

	Unit	Hours	Hours £	Mat'ls £	O & P £	Total £
15mm diameter, to timber	m	0.20	2.90	1.71	0.69	5.30
15mm diameter, plugged and screwed	m	0.22	3.19	1.97	0.77	5.93
extra over for						
reduced coupling, 15 × 8mm	nr	0.18	2.61	5.61	1.23	9.45
reduced coupling, 15 × 10mm	nr	0.18	2.61	2.21	0.72	5.54
reduced coupling, 15 × 12mm	nr	0.18	2.61	1.68	0.64	4.93
adaptor coupling, 15mm × 1/2in	nr	0.18	2.61	2.59	0.78	5.98
straight female connector, 15mm × 1/2in	nr	0.18	2.61	2.75	0.80	6.16
female reducing connector, 15mm × 3/8in	nr	0.18	2.61	5.42	1.20	9.23
female reducing connector, 15mm × 1/4in	nr	0.18	2.61	4.83	1.12	8.56
straight male connector, 15mm × 1/2in	nr	0.18	2.61	2.34	0.74	5.69

	Unit	Hours	Hours £	Mat'ls £	O & P £	Total £

15mm diameter pipe (cont'd)

	Unit	Hours	Hours £	Mat'ls £	O & P £	Total £
male reducing connector, 15mm × 3/4in	nr	0.18	2.61	3.16	0.87	6.64
tank connector, 15mm × 1/2in	nr	0.18	2.61	5.78	1.26	9.65
tank connector, 15mm × 3/4in	nr	0.18	2.61	5.78	1.26	9.65
reducer, 15 × 8mm	nr	0.18	2.61	1.86	0.67	5.14
reducer, 15 × 10mm	nr	0.18	2.61	0.73	0.50	3.84
reducer, 15 × 12mm	nr	0.18	2.61	1.46	0.61	4.68
female adaptor, 15 × 1/2in	nr	0.18	2.61	4.68	1.09	8.38
male adaptor, 15 × 1/2in	nr	0.18	2.61	4.78	1.11	8.50
adaptor, 15 × 1/2in	nr	0.18	2.61	2.27	0.73	5.61
male elbow, 15mm × 1/2in	nr	0.18	2.61	4.63	1.09	8.33
reducing male elbow, 15mm × 3/8in	nr	0.18	2.61	6.52	1.37	10.50
female elbow, 15mm × 1/2in	nr	0.18	2.61	4.00	0.99	7.60
reducing female elbow, 15mm × 1/4in	nr	0.18	2.61	6.38	1.35	10.34
backplate elbow, 15mm × 1/2in	nr	0.18	2.61	10.29	1.94	14.84
flanged bend, 15mm × 1/2in	nr	0.18	2.61	4.85	1.12	8.58
flanged bend, 15mm × 3/4in	nr	0.18	2.61	10.89	2.03	15.53
slow bend	nr	0.18	2.61	0.78	0.51	3.90
return bend	nr	0.18	2.61	7.27	1.48	11.36
obtuse elbow	nr	0.18	2.61	0.93	0.53	4.07
tee, reduced branch largest end 15mm	nr	0.22	3.19	8.77	1.79	13.75
tee, both ends reduced, largest end 15mm	nr	0.22	3.19	8.18	1.71	13.08
sweep tee, 90°, 15mm	nr	0.22	3.19	8.20	1.71	13.10
offset tee, 15mm	nr	0.22	3.19	10.84	2.10	16.13
double sweep tee, 15mm	nr	0.22	3.19	9.27	1.87	14.33
cross equal tee, 15mm	nr	0.22	3.19	12.27	2.32	17.78
stop end, 15mm	nr	0.18	2.61	3.17	0.87	6.65
tap connector, 15mm	nr	0.18	2.61	3.61	0.93	7.15
bent union adaptor, 15mm × 3/4in	nr	0.18	2.61	8.03	1.60	12.24

	Unit	Hours	Hours £	Mat'ls £	O & P £	Total £

15mm diameter pipe (cont'd)

	Unit	Hours	Hours £	Mat'ls £	O & P £	Total £
bent male union adaptor, 15mm × 1/2in	nr	0.18	2.61	9.45	1.81	13.87
bent female union adaptor, 15mm × 1/2in	nr	0.18	2.61	9.45	1.81	13.87
made bend	nr	0.18	2.61	0.00	0.39	3.00
22mm diameter, to timber	m	0.20	2.90	2.68	0.84	6.42
22mm diameter, plugged and screwed	m	0.22	3.19	2.91	0.92	7.02
extra over for						
straight coupling	nr	0.20	2.90	3.21	0.92	7.03
reduced coupling, 22 × 8mm	nr	0.20	2.90	3.21	0.92	7.03
reduced coupling, 22 × 10mm	nr	0.20	2.90	3.21	0.92	7.03
reduced coupling, 22 × 15mm	nr	0.20	2.90	2.16	0.76	5.82
straight female connector, 22mm × 3/4in	nr	0.20	2.90	3.94	1.03	7.87
female reducing connector, 22mm × 1/2in	· nr	0.20	2.90	6.26	1.37	10.53
straight male connector, 22mm × 3/4in	nr	0.20	2.90	4.12	1.05	8.07
male reducing connector, 22mm × 1/2in	nr	0.20	2.90	6.48	1.41	10.79
tank connector, 22mm × 1in	nr	0.20	2.90	8.81	1.76	13.47
reducer, 22 × 12mm	nr	0.20	2.90	3.33	0.93	7.16
female adaptor, 22 × 3/4in	nr	0.20	2.90	8.81	1.76	13.47
male adaptor, 22 × 3/4in	nr	0.20	2.90	6.10	1.35	10.35
adaptor, 22 × 3/4in	nr	0.20	2.90	2.51	0.81	6.22
street elbow, 22mm	nr	0.20	2.90	1.98	0.73	5.61
male elbow, 22mm × 3/4in	nr	0.20	2.90	6.57	1.42	10.89
female elbow, 22mm × 3/4in	nr	0.20	2.90	5.78	1.30	9.98
reducing female elbow, 22mm × 1/2in	nr	0.20	2.90	6.68	1.44	11.02
backplate elbow, 22mm × 3/4in	nr	0.20	2.90	10.42	2.00	15.32

	Unit	Hours	Hours £	Mat'ls £	O & P £	Total £
22mm diameter pipe (cont'd)						
flanged bend, 22mm × 3/4in	nr	0.20	2.90	11.99	2.23	17.12
overflow bend, 22mm × 3/4in	nr	0.20	2.90	14.69	2.64	20.23
slow bend	nr	0.20	2.90	1.53	0.66	5.09
return bend	nr	0.20	2.90	14.29	2.58	19.77
obtuse elbow	nr	0.20	2.90	1.94	0.73	5.57
tee, reduced branch largest end 22mm	nr	0.24	3.48	6.39	1.48	11.35
tee, both ends reduced, largest end 22mm	nr	0.24	3.48	2.76	0.94	7.18
corner tee, 22mm	nr	0.24	3.48	13.31	2.52	19.31
sweep tee, 90°, 22mm	nr	0.24	3.48	10.56	2.11	16.15
offset tee, 22mm	nr	0.24	3.48	13.05	2.48	19.01
double sweep tee, 22mm	nr	0.24	3.48	12.62	2.42	18.52
cross equal tee, 22mm	nr	0.24	3.48	13.71	2.58	19.77
stop end, 22mm	nr	0.20	2.90	2.23	0.77	5.90
tap connector, 22mm × 1/2in	nr	0.20	2.90	9.27	1.83	14.00
bent tap connector, 22mm × 1/2in	nr	0.20	2.90	11.30	2.13	16.33
bent tap connector, 22mm × 3/4in	nr	0.20	2.90	5.65	1.28	9.83
bent union adaptor, 22mm × 1in	nr	0.20	2.90	7.54	1.57	12.01
bent male union adaptor, 22mm × 3/4in	nr	0.20	2.90	9.28	1.83	14.01
bent female union adaptor, 22mm × 3/4in	nr	0.20	2.90	125.30	19.23	147.43
made bend	nr	0.20	2.90	0.00	0.44	3.34
Stop valves						
Gunmetal stop valve with brass headwork, copper × copper						
15mm	nr	0.20	2.90	5.03	1.19	9.12
22mm	nr	0.22	3.19	9.39	1.89	14.47
28mm	nr	0.28	4.06	26.67	4.61	35.34

	Unit	Hours	Hours £	Mat'ls £	O & P £	Total £
Gunmetal lockshield stop valve with brass headwork, copper × copper						
22mm	nr	0.22	3.19	14.50	2.65	20.34
28mm	nr	0.28	4.06	26.59	4.60	35.25
Valves						
Radiator valve, chromium plated, compression inlet × taper male union outlet						
8mm × 1/2in	nr	0.30	4.35	20.58	3.74	28.67
10mm × 1/2in	nr	0.30	4.35	22.62	4.05	31.02
Copper pipe, compression joints						
Copper pipe, compression joints and fittings, clips at maximum centres						
15mm diameter, to timber	m	0.20	2.90	1.64	0.68	5.22
15mm diameter, plugged and screwed	m	0.22	3.19	1.92	0.77	5.88
extra over for						
bent union radiator, 15mm × 1/2in	nr	0.18	2.61	6.54	1.37	10.52
air release elbow, 15mm	nr	0.18	2.61	8.44	1.66	12.71
slow bend, 15mm	nr	0.18	2.61	8.24	1.63	12.48
male elbow, 15mm × 1/2in	nr	0.18	2.61	2.29	0.74	5.64
male elbow, 15mm × 3/4in	nr	0.18	2.61	6.99	1.44	11.04
male elbow, BSP parallel thread, 15mm × 1/2in	nr	0.18	2.61	2.29	0.74	5.64
female elbow, 15mm × 1/2in	nr	0.18	2.61	3.52	0.92	7.05
offset tee, 15 × 15 × 15mm	nr	0.18	2.61	15.19	2.67	20.47
male adaptor, 15mm × 1/2in	nr	0.18	2.61	2.78	0.81	6.20
reducing set, 15 × 8mm	nr	0.18	2.61	1.89	0.68	5.18

	Unit	Hours	Hours £	Mat'ls £	O & P £	Total £
15mm diameter pipe (cont'd)						
reducing set, 15 × 10mm	nr	0.18	2.61	1.89	0.68	5.18
reducing set, 15 × 12mm	nr	0.18	2.61	1.89	0.68	5.18
22mm diameter, to timber	m	0.21	3.05	2.46	0.83	6.33
22mm diameter, plugged and screwed	m	0.23	3.34	2.64	0.90	6.87
extra over for						
air release elbow, 22mm	nr	0.18	2.61	11.17	2.07	15.85
slow bend, 22mm	nr	0.18	2.61	13.28	2.38	18.27
male elbow, 22mm × 3/4in	nr	0.18	2.61	2.96	0.84	6.41
male elbow, 22mm × 1in	nr	0.18	2.61	5.79	1.26	9.66
male elbow, BSP parallel thread, 22mm × 3/4in	nr	0.18	2.61	2.96	0.84	6.41
male elbow, BSP parallel thread, 22mm × 1in	nr	0.18	2.61	5.79	1.26	9.66
female elbow, 22mm × 3/4in	nr	0.18	2.61	5.09	1.16	8.86
female elbow, 22mm × 1in	nr	0.18	2.61	7.92	1.58	12.11
bent cylinder connector, 22mm × 1in	nr	0.18	2.61	11.64	2.14	16.39
tee with reduced branch, 22 × 22 × 15mm	nr	0.22	3.19	7.33	1.58	12.10
tee, end reduced, 22 × 15 × 22mm	nr	0.22	3.19	8.53	1.76	13.48
tee, end and branch reduced, 22 × 15 × 15mm	nr	0.22	3.19	8.23	1.71	13.13
straight swivel connector, 22mm × 1/2in	nr	0.18	2.61	6.87	1.42	10.90
bent swivel connector, 22mm × 1/2in	nr	0.18	2.61	9.08	1.75	13.44
offset tee, 22 × 22 × 15mm	nr	0.22	3.19	22.35	3.83	29.37

	Unit	Hours	Hours £	Mat'ls £	O & P £	Total £

Stop valves

Gunmetal stop valve with brass
headwork, copper × copper

	Unit	Hours	Hours £	Mat'ls £	O & P £	Total £
15mm	nr	0.19	2.76	6.90	1.45	11.10
22mm	nr	0.21	3.05	12.13	2.28	17.45
28mm	nr	0.27	3.92	31.63	5.33	40.88

Gunmetal lockshield stop valve
with brass headwork, copper ×
copper

	Unit	Hours	Hours £	Mat'ls £	O & P £	Total £
15mm	nr	0.19	2.76	15.00	2.66	20.42
22mm	nr	0.21	3.05	21.17	3.63	27.85

Cold water storage tanks

Galvanised steel cistern with
cover, reference

	Unit	Hours	Hours £	Mat'ls £	O & P £	Total £
SC10, 18 litres	nr	0.60	8.70	49.22	8.69	66.61
SC15, 36 litres	nr	0.65	9.43	52.02	9.22	70.66
SC20, 54 litres	nr	0.65	9.43	64.77	11.13	85.32
SC25, 68 litres	nr	0.70	10.15	68.26	11.76	90.17
SC30, 86 litres	nr	0.75	10.88	76.07	13.04	99.99
SC40, 114 litres	nr	0.80	11.60	82.17	14.07	107.84

Plastic cistern with lid, reference
cover, reference

	Unit	Hours	Hours £	Mat'ls £	O & P £	Total £
PC4, 18 litres	nr	0.60	8.70	6.82	2.33	17.85
PC15, 68 litres	nr	0.70	10.15	16.44	3.99	30.58
PC20, 91 litres	nr	0.75	10.88	16.75	4.14	31.77
PC25, 114 litres	nr	0.80	11.60	22.28	5.08	38.96
PC40, 182 litres	nr	0.90	13.05	39.83	7.93	60.81
PC50, 227 litres	nr	1.00	14.50	40.13	8.19	62.82

Cutting holes for connectors

	Unit	Hours	Hours £	Mat'ls £	O & P £	Total £
13mm	nr	0.10	1.45	0.00	0.22	1.67
18mm	nr	0.12	1.74	0.00	0.26	2.00
25mm	nr	0.15	2.18	0.00	0.33	2.50
32mm	nr	0.20	2.90	0.00	0.44	3.34

	Unit	Hours	Hours £	Mat'ls £	O & P £	Total £

Hot water copper cylinders

Direct cylinders, insulated,
reference

	Unit	Hours	Hours £	Mat'ls £	O & P £	Total £
ref 5, 98 litres	nr	0.50	7.25	124.34	19.74	151.33
ref 7, 120 litres	nr	0.55	7.98	141.93	22.49	172.39
ref 8, 148 litres	nr	0.65	9.43	156.08	24.83	190.33
ref 9, 166 litres	nr	0.85	12.33	170.37	27.40	210.10

Indirect cylinders, insulated,
reference

	Unit	Hours	Hours £	Mat'ls £	O & P £	Total £
ref 2, 96 litres	nr	0.50	7.25	98.48	15.86	121.59
ref 3, 114 litres	nr	0.55	7.98	107.11	17.26	132.35
ref 7, 117 litres	nr	0.55	7.98	112.62	18.09	138.68
ref 8, 140 litres	nr	0.65	9.43	128.41	20.68	158.51
ref 9, 162 litres	nr	0.70	10.15	159.80	25.49	195.44

Insulation

Preformed pipe lagging, fire-
retardant foam 13mm thick for
pipe size

	Unit	Hours	Hours £	Mat'ls £	O & P £	Total £
15mm	m	0.08	0.88	1.11	0.30	2.28
22mm	m	0.09	0.98	1.45	0.37	2.80
28mm	m	0.10	1.09	1.72	0.42	3.24

Preformed pipe lagging, fire-
retardant foam 25mm thick for
pipe size

	Unit	Hours	Hours £	Mat'ls £	O & P £	Total £
15mm	m	0.10	1.45	4.27	0.86	6.58
22mm	m	0.11	1.60	4.70	0.94	7.24
28mm	m	0.12	1.74	5.40	1.07	8.21

50mm glass fibre filled insulating
jacket, fixing bands to tank size

	Unit	Hours	Hours £	Mat'ls £	O & P £	Total £
445 × 305 × 300mm	nr	0.35	5.08	4.26	1.40	10.74
495 × 368 × 362mm	nr	0.40	5.80	5.21	1.65	12.66
630 × 450 × 420mm	nr	0.45	6.53	6.03	1.88	14.44
665 × 490 × 515mm	nr	0.50	7.25	7.07	2.15	16.47

	Unit	Hours	Hours £	Mat'ls £	O & P £	Total £

Boilers

Gas-fired wall-mounted
combination boiler for domestic
central heating and indirect hot
water

Fan flue, output

	Unit	Hours	Hours £	Mat'ls £	O & P £	Total £
30,400-81,900 BTU	nr	4.80	69.60	651.68	108.19	829.47
37,500-81,900 BTU	nr	4.80	69.60	696.68	114.94	881.22

Condensing fan flue, output

	Unit	Hours	Hours £	Mat'ls £	O & P £	Total £
25,300-92,800 BTU	nr	4.80	69.60	815.68	132.79	1018.07
25,900-93,800 BTU	nr	4.80	69.60	843.68	136.99	1050.27

Oil storage tanks

Standard domestic pattern plastic
oil storage tank

	Unit	Hours	Hours £	Mat'ls £	O & P £	Total £
1380 × 1780 × 890mm capacity 1200 litres	nr	1.50	21.75	170.48	28.83	221.06
745 × 1700 × 1380mm capacity 1225 litres	nr	1.50	21.75	241.93	39.55	303.23
1190mm diameter × 1400 high, capacity 1300 litres	nr	1.50	21.75	170.86	28.89	221.50
1130 × 2055 × 1145mm capacity 1800 litres	nr	1.75	25.38	299.49	48.73	373.59
1265 × 2250 × 1320mm capacity 2500 litres	nr	2.00	29.00	373.05	60.31	462.36
1640mm diameter × 1750 high, capacity 3600 litres	nr	2.50	36.25	411.42	67.15	514.82
2480mm diameter × 2640 high, capacity 10,000 litres	nr	3.00	43.50	1189.64	184.97	1418.11

	Unit	Hours	Hours £	Mat'ls £	O & P £	Total £
Standard domestic pattern plastic bunded oil storage tank						
1380 × 1780 × 890mm capacity						
1200 litres	nr	1.50	21.75	228.27	37.50	287.52
745 × 1700 × 1380mm capacity						
1225 litres	nr	1.50	21.75	299.34	48.16	369.25
1190mm diameter × 1400 high,						
capacity 1300 litres	nr	1.50	21.75	699.75	108.23	829.73
1130 × 2055 × 1145mm capacity						
1800 litres	nr	1.75	25.38	730.58	113.39	869.35
1265 × 2250 × 1320mm capacity						
2500 litres	nr	2.00	29.00	811.36	126.05	966.41

Radiators

**Single convector radiator
fixed to concealed brackets
plugged and screwed to brickwork**

	Unit	Hours	Hours £	Mat'ls £	O & P £	Total £
450mm high, length						
500mm	nr	1.00	14.50	14.43	4.34	33.27
1000mm	nr	1.15	16.68	25.19	6.28	48.14
1400mm	nr	1.25	18.13	33.62	7.76	59.51
1800mm	nr	1.40	20.30	48.53	10.32	79.15
2000mm	nr	1.50	21.75	64.56	12.95	99.26
600mm high, length						
500mm	nr	1.10	15.95	17.80	5.06	38.81
1000mm	nr	1.25	18.13	30.60	7.31	56.03
1400mm	nr	1.35	19.58	41.64	9.18	70.40
1800mm	nr	1.50	21.75	60.58	12.35	94.68
2000mm	nr	1.60	23.20	80.54	15.56	119.30
700mm high, length						
500mm	nr	1.20	17.40	19.80	5.58	42.78
1000mm	nr	1.35	19.58	34.60	8.13	62.30
1400mm	nr	1.45	21.03	63.64	12.70	97.36
1800mm	nr	1.60	23.20	82.58	15.87	121.65
2000mm	nr	1.70	24.65	91.54	17.43	133.62

	Unit	Hours	Hours £	Mat'ls £	O & P £	Total £

Double convector radiator
fixed to concealed brackets
plugged and screwed to brickwork

450mm high, length

	Unit	Hours	Hours £	Mat'ls £	O & P £	Total £
500mm	nr	1.00	14.50	23.43	5.69	43.62
1000mm	nr	1.15	16.68	43.19	8.98	68.84
1400mm	nr	1.25	18.13	66.62	12.71	97.46
2000mm	nr	1.40	20.30	107.53	19.17	147.00
2600mm	nr	1.50	21.75	118.56	21.05	161.36

600mm high, length

	Unit	Hours	Hours £	Mat'ls £	O & P £	Total £
500mm	nr	1.10	15.95	28.10	6.61	50.66
1000mm	nr	1.25	18.13	53.17	10.69	81.99
1400mm	nr	1.35	19.58	82.69	15.34	117.60
2000mm	nr	1.50	21.75	133.28	23.25	178.28
2600mm	nr	1.60	23.20	147.96	25.67	196.83

Double panel radiator
 fixed to concealed brackets
plugged and screwed to
brickwork

450mm high, length

	Unit	Hours	Hours £	Mat'ls £	O & P £	Total £
500mm	nr	1.00	14.50	19.43	5.09	39.02
1000mm	nr	1.15	16.68	34.19	7.63	58.49
1400mm	nr	1.25	18.13	52.62	10.61	81.36
2000mm	nr	1.40	20.30	84.53	15.72	120.55
2600mm	nr	1.50	21.75	93.56	17.30	132.61

600mm high, length

	Unit	Hours	Hours £	Mat'ls £	O & P £	Total £
500mm	nr	1.10	15.95	23.10	5.86	44.91
1000mm	nr	1.25	18.13	42.19	9.05	69.36
1400mm	nr	1.35	19.58	65.68	12.79	98.04
2000mm	nr	1.50	21.75	105.28	19.05	146.08
2600mm	nr	1.60	17.50	116.96	20.17	154.63

	Unit	Hours	Hours £	Mat'ls £	O & P £	Total £

GLAZING

Clear float glass

In wood with putty and sprigs

under 0.15m2, thickness

3mm	m2	0.90	10.80	25.82	5.49	42.11
4mm	m2	0.90	10.80	27.58	5.76	44.14
5mm	m2	0.90	10.80	31.75	6.38	48.93
6mm	m2	1.00	12.00	33.60	6.84	52.44
10mm	m2	1.05	12.60	57.60	10.53	80.73

over 0.15m2, thickness

3mm	m2	0.60	7.20	25.82	4.95	37.97
4mm	m2	0.60	7.20	27.58	5.22	40.00
5mm	m2	0.60	7.20	31.75	5.84	44.79
6mm	m2	0.65	7.80	33.60	6.21	47.61
10mm	m2	0.70	8.40	57.60	9.90	75.90

In wood with pinned beads

under 0.15m2, thickness

3mm	m2	1.10	13.20	25.82	5.85	44.87
4mm	m2	1.10	13.20	27.58	6.12	46.90
5mm	m2	1.10	13.20	31.75	6.74	51.69
6mm	m2	1.20	14.40	33.60	7.20	55.20
10mm	m2	1.25	15.00	57.60	10.89	83.49

over 0.15m2, thickness

3mm	m2	0.80	9.60	25.82	5.31	40.73
4mm	m2	0.80	9.60	27.58	5.58	42.76
5mm	m2	0.80	9.60	31.75	6.20	47.55
6mm	m2	0.90	10.80	33.60	6.66	51.06
10mm	m2	0.95	11.40	57.60	10.35	79.35

	Unit	Hours	Hours £	Mat'ls £	O & P £	Total £

In wood with screwed beads

under 0.15m2, thickness

3mm	m2	1.30	15.60	25.82	6.21	47.63
4mm	m2	1.30	15.60	27.58	6.48	49.66
5mm	m2	1.30	15.60	31.75	7.10	54.45
6mm	m2	1.40	16.80	33.60	7.56	57.96
10mm	m2	1.45	17.40	57.60	11.25	86.25

over 0.15m2, thickness

3mm	m2	1.00	12.00	25.82	5.67	43.49
4mm	m2	1.00	12.00	27.58	5.94	45.52
5mm	m2	1.00	12.00	31.75	6.56	50.31
6mm	m2	1.05	12.60	33.60	6.93	53.13
10mm	m2	1.10	13.20	57.60	10.62	81.42

In metal with putty

under 0.15m2, thickness

3mm	m2	0.90	10.80	25.82	5.49	42.11
4mm	m2	0.90	10.80	27.58	5.76	44.14
5mm	m2	0.90	10.80	31.75	6.38	48.93
6mm	m2	1.00	12.00	33.60	6.84	52.44
10mm	m2	1.05	12.60	57.60	10.53	80.73

over 0.15m2, thickness

3mm	m2	0.60	7.20	25.82	4.95	37.97
4mm	m2	0.60	7.20	27.58	5.22	40.00
5mm	m2	0.60	7.20	31.75	5.84	44.79
6mm	m2	0.65	7.80	33.60	6.21	47.61
10mm	m2	0.70	8.40	57.60	9.90	75.90

In metal with clips and putty

under 0.15m2, thickness

3mm	m2	0.95	11.40	25.82	5.58	42.80
4mm	m2	0.95	11.40	27.58	5.85	44.83
5mm	m2	0.95	11.40	31.75	6.47	49.62
6mm	m2	1.05	12.60	33.60	6.93	53.13
10mm	m2	1.10	13.20	57.60	10.62	81.42

	Unit	Hours	Hours £	Mat'ls £	O & P £	Total £
over 0.15m2, thickness						
3mm	m2	0.65	7.80	25.82	5.04	38.66
4mm	m2	0.65	7.80	27.58	5.31	40.69
5mm	m2	0.65	7.80	31.75	5.93	45.48
6mm	m2	0.70	8.40	33.60	6.30	48.30
10mm	m2	0.75	9.00	57.60	9.99	76.59

In metal with screwed metal beads

	Unit	Hours	Hours £	Mat'ls £	O & P £	Total £
under 0.15m2, thickness						
3mm	m2	1.30	15.60	25.82	6.21	47.63
4mm	m2	1.30	15.60	27.58	6.48	49.66
5mm	m2	1.30	15.60	31.75	7.10	54.45
6mm	m2	1.40	16.80	33.60	7.56	57.96
10mm	m2	1.45	17.40	57.60	11.25	86.25
over 0.15m2, thickness						
3mm	m2	1.00	12.00	25.82	5.67	43.49
4mm	m2	1.00	12.00	27.58	5.94	45.52
5mm	m2	1.00	12.00	31.75	6.56	50.31
6mm	m2	1.05	12.60	33.60	6.93	53.13
10mm	m2	1.10	13.20	57.60	10.62	81.42

White patterned glass

In wood with putty and sprigs

	Unit	Hours	Hours £	Mat'ls £	O & P £	Total £
under 0.15m2, thickness						
4mm	m2	0.90	10.80	30.41	6.18	47.39
6mm	m2	1.00	12.00	35.68	7.15	54.83
over 0.15m2, thickness						
4mm	m2	0.60	7.20	30.41	5.64	43.25
6mm	m2	0.65	7.80	35.68	6.52	50.00

In wood with pinned beads

	Unit	Hours	Hours £	Mat'ls £	O & P £	Total £
under 0.15m2, thickness						
4mm	m2	1.10	13.20	30.41	6.54	50.15
6mm	m2	1.20	14.40	35.68	7.51	57.59

	Unit	Hours	Hours £	Mat'ls £	O & P £	Total £
over 0.15m2, thickness						
4mm	m2	0.80	9.60	30.41	6.00	46.01
6mm	m2	0.85	10.20	35.68	6.88	52.76
In wood with screwed beads						
under 0.15m2, thickness						
4mm	m2	1.30	15.60	30.41	6.90	52.91
6mm	m2	1.40	16.80	35.68	7.87	60.35
over 0.15m2, thickness						
4mm	m2	1.00	12.00	30.41	6.36	48.77
6mm	m2	1.50	18.00	35.68	8.05	61.73
In metal with putty						
under 0.15m2, thickness						
4mm	m2	0.90	10.80	30.41	6.18	47.39
6mm	m2	1.00	12.00	35.68	7.15	54.83
over 0.15m2, thickness						
4mm	m2	0.60	7.20	30.41	5.64	43.25
6mm	m2	0.65	7.80	35.68	6.52	50.00
In metal with clips and putty						
under 0.15m2, thickness						
4mm	m2	0.95	11.40	30.41	6.27	48.08
6mm	m2	1.05	12.60	35.68	7.24	55.52
over 0.15m2, thickness						
4mm	m2	0.65	7.80	30.41	5.73	43.94
6mm	m2	0.70	8.40	35.68	6.61	50.69
In metal with screwed metal beads						
under 0.15m2, thickness						
4mm	m2	1.30	15.60	30.41	6.90	52.91
6mm	m2	1.40	16.80	35.68	7.87	60.35

	Unit	Hours	Hours £	Mat'ls £	O & P £	Total £
over 0.15m2, thickness						
4mm	m2	1.00	12.00	30.41	6.36	48.77
6mm	m2	1.05	12.60	35.68	7.24	55.52

Georgian wired cast glass

In wood with putty and sprigs

under 0.15m2, thickness						
7mm	m2	0.90	10.80	31.65	6.37	48.82
over 0.15m2, thickness						
7mm	m2	0.65	7.80	31.65	5.92	45.37

In wood with pinned beads

under 0.15m2, thickness						
7mm	m2	1.20	14.40	31.65	6.91	52.96
over 0.15m2, thickness						
7mm	m2	0.90	10.80	31.65	6.37	48.82

In wood with screwed beads

under 0.15m2, thickness						
7mm	m2	1.40	16.80	31.65	7.27	55.72
over 0.15m2, thickness						
7mm	m2	1.05	12.60	31.65	6.64	50.89

In metal with putty and clips

under 0.15m2, thickness						
7mm	m2	1.05	12.60	31.65	6.64	50.89
over 0.15m2, thickness						
7mm	m2	0.70	8.40	31.65	6.01	46.06

	Unit	Hours	Hours £	Mat'ls £	O & P £	Total £
In metal with screwed metal beads						
under 0.15m2, thickness 7mm	m2	1.40	16.80	31.65	7.27	55.72
over 0.15m2, thickness 7mm	m2	1.05	12.60	31.65	6.64	50.89
Georgian wired polished glass						
In wood with putty and sprigs						
under 0.15m2, thickness 6mm	m2	0.90	10.80	74.67	12.82	98.29
over 0.15m2, thickness 6mm	m2	0.65	7.80	74.67	12.37	94.84
In wood with pinned beads						
under 0.15m2, thickness 6mm	m2	1.20	14.40	74.67	13.36	102.43
over 0.15m2, thickness 6mm	m2	0.90	10.80	74.67	12.82	98.29
In wood with screwed beads						
under 0.15m2, thickness 6mm	m2	1.40	16.80	74.67	13.72	105.19
over 0.15m2, thickness 6mm	m2	1.05	12.60	74.67	13.09	100.36

	Unit	Hours	Hours £	Mat'ls £	O & P £	Total £
In metal with putty and clips						
under 0.15m2, thickness						
7mm	m2	1.05	12.60	74.67	13.09	100.36
over 0.15m2, thickness						
7mm	m2	0.70	8.40	74.67	12.46	95.53
In metal with screwed metal beads						
under 0.15m2, thickness						
7mm	m2	1.40	16.80	74.67	13.72	105.19
over 0.15m2, thickness						
7mm	m2	1.05	12.60	74.67	13.09	100.36

Clear laminated safety glass

In wood with pinned beads

	Unit	Hours	Hours £	Mat'ls £	O & P £	Total £
under 0.15m2, thickness						
4mm	m2	1.10	13.20	53.40	9.99	76.59
4.4mm	m2	1.10	13.20	59.69	10.93	83.82
6.4mm	m2	1.20	14.40	64.53	11.84	90.77
over 0.15m2, thickness						
4mm	m2	0.80	9.60	53.40	9.45	72.45
4.4mm	m2	0.80	9.60	59.69	10.39	79.68
6.4mm	m2	0.90	10.80	64.53	11.30	86.63

In wood with screwed beads

	Unit	Hours	Hours £	Mat'ls £	O & P £	Total £
under 0.15m2, thickness						
4mm	m2	1.30	15.60	53.40	10.35	79.35
4.4mm	m2	1.30	15.60	59.69	11.29	86.58
6.4mm	m2	1.40	16.80	64.53	12.20	93.53

	Unit	Hours	Hours £	Mat'ls £	O & P £	Total £
over 0.15m2, thickness						
4mm	m2	1.00	12.00	53.40	9.81	75.21
4.4mm	m2	1.00	12.00	59.69	10.75	82.44
6.4mm	m2	1.05	12.60	64.53	11.57	88.70
In metal with screwed beads						
under 0.15m2, thickness						
4mm	m2	1.30	15.60	53.40	10.35	79.35
4.4mm	m2	1.30	15.60	59.69	11.29	86.58
6.4mm	m2	1.40	16.80	64.53	12.20	93.53
over 0.15m2, thickness						
4mm	m2	1.00	12.00	53.40	9.81	75.21
4.4mm	m2	1.00	12.00	59.69	10.75	82.44
6.4mm	m2	1.05	12.60	64.53	11.57	88.70

	Unit	Hours	Hours £	Mat'ls £	O & P £	Total £

PAINTING AND WALLPAPERING

On internal surfaces before fixing

One coat of primer on wood
surfaces

exceeding 300mm girth	m2	0.16	1.92	0.52	0.37	2.81
not exceeding 150mm girth	m	0.04	0.48	0.09	0.09	0.66
150-300mm girth	m	0.06	0.72	0.17	0.13	1.02
isolated areas not exceeding 0.5m2	nr	0.10	1.20	0.26	0.22	1.68

One coat of primer on metal
surfaces

exceeding 300mm girth	m2	0.16	1.92	0.67	0.39	2.98
not exceeding 150mm girth	m	0.04	0.48	0.11	0.09	0.68
150-300mm girth	m	0.06	0.72	0.21	0.14	1.07
isolated areas not exceeding 0.5m2	nr	0.10	1.20	0.33	0.23	1.76

One coat of wood preservative
on wood surfaces

exceeding 300mm girth	m2	0.14	1.68	0.34	0.30	2.32
not exceeding 150mm girth	m	0.03	0.36	0.09	0.07	0.52
150-300mm girth	m	0.05	0.60	0.13	0.11	0.84
isolated areas not exceeding 0.5m2	nr	0.08	0.96	0.17	0.17	1.30

On internal walls and ceilings

One coat of emulsion paint on

brickwork or blockwork walls	m2	0.20	2.40	0.65	0.46	3.51
brickwork or blockwork walls in staircase areas	m2	0.26	3.12	0.65	0.57	4.34
plastered walls	m2	0.12	1.44	0.43	0.28	2.15
plastered walls in staircase areas	m2	0.16	1.92	0.43	0.35	2.70

	Unit	Hours	Hours £	Mat'ls £	O & P £	Total £
plastered ceilings	m2	0.20	2.40	0.43	0.42	3.25
plastered ceilings in staircase areas	m2	0.24	2.88	0.43	0.50	3.81
embossed papered walls	m2	0.16	1.92	0.54	0.37	2.83
embossed papered walls in staircase areas	m2	0.20	2.40	0.54	0.44	3.38

Two coats of emulsion paint on

	Unit	Hours	Hours £	Mat'ls £	O & P £	Total £
brickwork or blockwork walls	m2	0.28	3.36	1.30	0.70	5.36
brickwork or blockwork walls in staircase areas	m2	0.32	3.84	1.30	0.77	5.91
plastered walls	m2	0.16	1.92	0.90	0.42	3.24
plastered walls in staircase areas	m2	0.20	2.40	0.90	0.50	3.80
plastered ceilings	m2	0.24	2.88	0.90	0.57	4.35
plastered ceilings in staircase areas	m2	0.28	3.36	0.90	0.64	4.90
embossed papered walls	m2	0.24	2.88	1.09	0.60	4.57
embossed papered walls in staircase areas	m2	0.26	3.12	1.09	0.63	4.84

One mist coat and two coats of emulsion paint on

	Unit	Hours	Hours £	Mat'ls £	O & P £	Total £
brickwork or blockwork walls	m2	0.36	4.32	1.63	0.89	6.84
brickwork or blockwork walls in staircase areas	m2	0.40	4.80	1.63	0.96	7.39
plastered walls	m2	0.24	2.88	1.09	0.60	4.57
plastered walls in staircase areas	m2	0.28	3.36	1.09	0.67	5.12
plastered ceilings	m2	0.32	3.84	1.09	0.74	5.67
plastered ceilings in staircase areas	m2	0.36	4.32	1.09	0.81	6.22
embossed papered walls	m2	0.30	3.60	1.36	0.74	5.70
embossed papered walls in staircase areas	m2	0.34	4.08	1.36	0.82	6.26

	Unit	Hours	Hours £	Mat'ls £	O & P £	Total £

One coat sealer primer, one
undercoat and one coat gloss
on

plastered walls	m2	0.42	5.04	1.30	0.95	7.29
plastered walls in staircase areas	m2	0.46	5.52	1.30	1.02	7.84
plastered ceilings	m2	0.48	5.76	1.30	1.06	8.12
plastered ceilings in staircase areas	m2	0.52	6.24	1.30	1.13	8.67

On internal woodwork

One coat primer on wood
general surfaces

exceeding 300mm girth	m2	0.18	2.16	0.54	0.41	3.11
not exceeding 150mm girth	m	0.04	0.48	0.09	0.09	0.66
150-300mm girth	m	0.06	0.72	0.17	0.13	1.02
isolated areas not exceeding 0.5m2	nr	0.10	1.20	0.26	0.22	1.68

One coat primer on wood
windows, glazed doors and the
like

panes area not exceeding 0.1m2	m2	0.35	4.20	0.19	0.66	5.05
panes area 0.1 to 0.5m2	m2	0.30	3.60	0.17	0.57	4.34
panes area 0.5 to 1m2	m2	0.25	3.00	0.15	0.47	3.62
panes area exceeding 1m2	m2	0.22	2.64	0.13	0.42	3.19

One coat primer on wood frames,
linings, skirtings and the like

exceeding 300mm girth	m2	0.20	2.40	0.50	0.44	3.34
not exceeding 150mm girth	m	0.05	0.60	0.07	0.10	0.77
150-300mm girth	m	0.08	0.96	0.15	0.17	1.28
isolated areas not exceeding 0.5m2	nr	0.11	1.32	0.23	0.23	1.78

	Unit	Hours	Hours £	Mat'ls £	O & P £	Total £
One coat undercoat on primed wood general surfaces						
exceeding 300mm girth	m2	0.18	2.16	0.54	0.41	3.11
not exceeding 150mm girth	m	0.04	0.48	0.09	0.09	0.66
150-300mm girth	m	0.06	0.72	0.17	0.13	1.02
isolated areas not exceeding 0.5m2	nr	0.10	1.20	0.26	0.22	1.68
One coat undercoat on primed wood windows, glazed doors and the like						
panes area not exceeding 0.1m2	m2	0.35	4.20	0.21	0.66	5.07
panes area 0.1 to 0.5m2	m2	0.30	3.60	0.19	0.57	4.36
panes area 0.5 to 1m2	m2	0.25	3.00	0.17	0.48	3.65
panes area exceeding 1m2	m2	0.22	2.64	0.15	0.42	3.21
One coat undercoat on primed wood frames, linings, skirtings and the like						
exceeding 300mm girth	m2	0.20	2.40	0.54	0.44	3.38
not exceeding 150mm girth	m	0.05	0.60	0.09	0.10	0.79
150-300mm girth	m	0.08	0.96	0.17	0.17	1.30
isolated areas not exceeding 0.5m2	nr	0.11	1.32	0.26	0.24	1.82
Two coats undercoat on primed wood general surfaces						
exceeding 300mm girth	m2	0.36	4.32	1.09	0.81	6.22
not exceeding 150mm girth	m	0.08	0.96	0.17	0.17	1.30
150-300mm girth	m	0.12	1.44	0.34	0.27	2.05
isolated areas not exceeding 0.5m2	nr	0.20	2.40	0.52	0.44	3.36

	Unit	Hours	Hours £	Mat'ls £	O & P £	Total £
Two coats undercoat on primed wood windows, glazed doors and the like						
panes area not exceeding 0.1m2	m2	0.60	7.20	0.43	1.14	8.77
panes area 0.1 to 0.5m2	m2	0.50	6.00	0.39	0.96	7.35
panes area 0.5 to 1m2	m2	0.40	4.80	0.34	0.77	5.91
panes area exceeding 1m2	m2	0.30	3.60	0.30	0.59	4.49
Two coats undercoat on primed wood frames, linings, skirtings and the like						
exceeding 300mm girth	m2	0.40	4.80	12.09	2.53	19.42
not exceeding 150mm girth	m	0.10	1.20	0.17	0.21	1.58
150-300mm girth	m	0.16	1.92	0.34	0.34	2.60
isolated areas not exceeding 0.5m2	nr	0.22	2.65	0.52	0.48	3.65
One coat gloss on undercoated wood general surfaces						
exceeding 300mm girth	m2	0.18	2.16	0.54	0.41	3.11
not exceeding 150mm girth	m	0.04	0.48	0.09	0.09	0.66
150-300mm girth	m	0.06	0.72	0.17	0.13	1.02
isolated areas not exceeding 0.5m2	nr	0.10	1.20	0.26	0.22	1.68
One coat gloss on undercoated wood windows, glazed doors and the like						
panes area not exceeding 0.1m2	m2	0.35	4.20	0.21	0.66	5.07
panes area 0.1 to 0.5m2	m2	0.30	3.60	0.19	0.57	4.36
panes area 0.5 to 1m2	m2	0.25	3.00	0.17	0.48	3.65
panes area exceeding 1m2	m2	0.22	2.64	0.15	0.42	3.21

	Unit	Hours	Hours £	Mat'ls £	O & P £	Total £
One coat gloss on undercoated wood frames, linings, skirtings and the like						
exceeding 300mm girth	m2	0.20	2.40	0.54	0.44	3.38
not exceeding 150mm girth	m	0.05	0.60	0.09	0.10	0.79
150-300mm girth	m	0.08	0.96	0.17	0.17	1.30
isolated areas not exceeding 0.5m2	nr	0.11	1.32	0.26	0.24	1.82
One coat primer, one coat undercoat and one coat gloss on wood general surfaces						
exceeding 300mm girth	m2	0.54	6.48	1.63	1.22	9.33
not exceeding 150mm girth	m	0.12	1.44	0.26	0.26	1.96
150-300mm girth	m	0.18	2.16	0.52	0.40	3.08
isolated areas not exceeding 0.5m2	nr	0.30	3.60	0.78	0.66	5.04
One coat primer, one coat undercoat and one coat gloss on wood windows, glazed doors and the like						
panes area not exceeding 0.1m2	m2	0.75	9.00	0.63	1.44	11.07
panes area 0.1 to 0.5m2	m2	0.60	7.20	0.47	1.15	8.82
panes area 0.5 to 1m2	m2	0.50	6.00	0.50	0.98	7.48
panes area exceeding 1m2	m2	0.40	4.80	0.43	0.78	6.01
One coat primer, one coat undercoat and one coat gloss on wood frames, linings, skirtings and the like						
exceeding 300mm girth	m2	0.60	7.20	1.59	1.32	10.11
not exceeding 150mm girth	m	0.15	1.80	0.23	0.30	2.33
150-300mm girth	m	0.24	2.88	0.50	0.51	3.89
isolated areas not exceeding 0.5m2	nr	0.33	3.96	0.76	0.71	5.43

	Unit	Hours	Hours £	Mat'ls £	O & P £	Total £
One coat primer, two coats undercoat and one coat gloss on wood general surfaces						
exceeding 300mm girth	m2	0.72	8.64	2.18	1.62	12.44
not exceeding 150mm girth	m	0.16	1.92	0.34	0.34	2.60
150-300mm girth	m	0.24	2.88	0.69	0.54	4.11
isolated areas not exceeding 0.5m2	nr	0.40	4.80	1.04	0.88	6.72
One coat primer, two coats undercoat and one coat gloss on wood windows, glazed doors and the like						
panes area not exceeding 0.1m2	m2	1.00	12.00	0.85	1.93	14.78
panes area 0.1 to 0.5m2	m2	0.80	9.60	0.76	1.55	11.91
panes area 0.5 to 1m2	m2	0.70	8.40	0.67	1.36	10.43
panes area exceeding 1m2	m2	0.60	7.20	0.58	1.17	8.95
One coat primer, two coats undercoat and one coat gloss on wood frames, linings, skirtings and the like						
exceeding 300mm girth	m2	0.80	9.60	2.13	1.76	13.49
not exceeding 150mm girth	m	0.20	2.40	0.32	0.41	3.13
150-300mm girth	m	0.32	3.84	0.67	0.68	5.19
isolated areas not exceeding 0.5m2	nr	0.44	5.28	1.02	0.95	7.25

	Unit	Hours	Hours £	Mat'ls £	O & P £	Total £

On internal metalwork

One coat primer on metal
general surfaces

	Unit	Hours	Hours £	Mat'ls £	O & P £	Total £
exceeding 300mm girth	m2	0.18	2.16	0.59	0.41	3.16
not exceeding 150mm girth	m	0.04	0.48	0.11	0.09	0.68
150-300mm girth	m	0.06	0.72	0.19	0.14	1.05
isolated areas not exceeding 0.5m2	nr	0.10	1.20	0.30	0.23	1.73

One coat primer on metal
windows, glazed doors and the
like

	Unit	Hours	Hours £	Mat'ls £	O & P £	Total £
panes area not exceeding 0.1m2	m2	0.35	4.20	0.21	0.66	5.07
panes area 0.1 to 0.5m2	m2	0.30	3.60	0.19	0.57	4.36
panes area 0.5 to 1m2	m2	0.25	3.00	0.17	0.48	3.65
panes area exceeding 1m2	m2	0.22	2.64	0.15	0.42	3.21

One coat primer on structural
steelwork

	Unit	Hours	Hours £	Mat'ls £	O & P £	Total £
exceeding 300mm girth	m2	0.24	2.88	0.59	0.52	3.99

One coat primer on metal
radiators

	Unit	Hours	Hours £	Mat'ls £	O & P £	Total £
exceeding 300mm girth	m2	0.18	2.16	0.59	0.41	3.16

One coat undercoat on metal
general surfaces

	Unit	Hours	Hours £	Mat'ls £	O & P £	Total £
exceeding 300mm girth	m2	0.18	2.16	0.54	0.41	3.11
not exceeding 150mm girth	m	0.04	0.48	0.09	0.09	0.66
150-300mm girth	m	0.06	0.72	0.17	0.13	1.02
isolated areas not exceeding 0.5m2	nr	0.10	1.20	0.26	0.22	1.68

	Unit	Hours	Hours £	Mat'ls £	O & P £	Total £
One coat undercoat on metal windows, glazed doors and the like						
panes area not exceeding 0.1m2	m2	0.35	4.20	0.19	0.66	5.05
panes area 0.1 to 0.5m2	m2	0.30	3.60	0.17	0.57	4.34
panes area 0.5 to 1m2	m2	0.25	3.00	0.15	0.47	3.62
panes area exceeding 1m2	m2	0.22	2.64	0.13	0.42	3.19
One coat undercoat on structural steelwork						
exceeding 300mm girth	m2	0.24	2.88	0.54	0.51	3.93
One coat undercoat on metal radiators						
exceeding 300mm girth	m2	0.18	2.16	0.54	0.41	3.11
Two coats undercoat on metal general surfaces						
exceeding 300mm girth	m2	0.18	2.16	1.09	0.49	3.74
not exceeding 150mm girth	m	0.04	0.48	0.17	0.10	0.75
150-300mm girth	m	0.06	0.72	0.34	0.16	1.22
isolated areas not exceeding 0.5m2	nr	0.10	1.20	0.52	0.26	1.98
Two coats undercoat on metal windows, glazed doors and the like						
panes area not exceeding 0.1m2	m2	0.35	4.20	0.39	0.69	5.28
panes area 0.1 to 0.5m2	m2	0.30	3.60	0.34	0.59	4.53
panes area 0.5 to 1m2	m2	0.25	3.00	0.30	0.50	3.80
panes area exceeding 1m2	m2	0.22	2.64	0.26	0.44	3.34

	Unit	Hours	Hours £	Mat'ls £	O & P £	Total £
Two coats undercoat on structural steelwork						
exceeding 300mm girth	m2	0.24	2.88	1.09	0.60	4.57
Two coats undercoat on metal radiators						
exceeding 300mm girth	m2	0.18	2.16	1.09	0.49	3.74
One coat gloss on metal general surfaces						
exceeding 300mm girth	m2	0.18	2.16	0.54	0.41	3.11
not exceeding 150mm girth	m	0.04	0.48	0.09	0.09	0.66
150-300mm girth	m	0.06	0.72	0.17	0.13	1.02
isolated areas not exceeding 0.5m2	nr	0.10	1.20	0.26	0.22	1.68
One coat gloss on metal windows, glazed doors and the like						
panes area not exceeding 0.1m2	m2	0.35	4.20	0.19	0.66	5.05
panes area 0.1 to 0.5m2	m2	0.30	3.60	0.17	0.57	4.34
panes area 0.5 to 1m2	m2	0.25	3.00	0.15	0.47	3.62
panes area exceeding 1m2	m2	0.22	2.64	0.13	0.42	3.19
One coat gloss on structural steelwork						
exceeding 300mm girth	m2	0.24	2.88	0.54	0.51	3.93
One coat gloss on metal radiators						
exceeding 300mm girth	m2	0.18	2.16	0.54	0.41	3.11

	Unit	Hours	Hours £	Mat'ls £	O & P £	Total £
One coat primer, one coat undercoat and one coat gloss on metal general surfaces						
exceeding 300mm girth	m2	0.54	6.48	1.68	1.22	9.38
not exceeding 150mm girth	m	0.12	1.44	0.28	0.26	1.98
150-300mm girth	m	0.18	2.16	0.54	0.41	3.11
isolated areas not exceeding 0.5m2	nr	0.30	3.60	0.82	0.66	5.08
One coat primer, one coat undercoat and one coat gloss on wood windows, glazed doors and the like						
panes area not exceeding 0.1m2	m2	0.75	9.00	0.65	1.45	11.10
panes area 0.1 to 0.5m2	m2	0.60	7.20	0.50	1.16	8.86
panes area 0.5 to 1m2	m2	0.50	6.00	0.52	0.98	7.50
panes area exceeding 1m2	m2	0.40	4.80	0.45	0.79	6.04
One coat primer, one coat undercoat and one coat gloss on structural steelwork						
exceeding 300mm girth	m2	0.60	7.20	1.68	1.33	10.21
One coat primer, one coat undercoat and one coat gloss on metal radiators						
exceeding 300mm girth	m2	0.52	6.24	1.68	1.19	9.11
One coat primer, two coats undercoat and one coat gloss on metal general surfaces						
exceeding 300mm girth	m2	0.54	6.48	2.23	1.31	10.02
not exceeding 150mm girth	m	0.12	1.44	0.28	0.26	1.98
150-300mm girth	m	0.18	2.16	0.54	0.41	3.11
isolated areas not exceeding .5m2	nr	0.30	3.60	0.80	0.66	5.06

	Unit	Hours	Hours £	Mat'ls £	O & P £	Total £

One coat primer, two coats
undercoat and one coat gloss on
metal windows, glazed doors
and the like

	Unit	Hours	Hours £	Mat'ls £	O & P £	Total £
panes area not exceeding 0.1m2	m2	0.75	9.00	0.80	1.47	11.27
panes area 0.1 to 0.5m2	m2	0.60	7.20	0.71	1.19	9.10
panes area 0.5 to 1m2	m2	0.50	6.00	0.65	1.00	7.65
panes area exceeding 1m2	m2	0.40	4.80	0.54	0.80	6.14

One coat primer, two coats
undercoat and one coat gloss on
structural steelwork

	Unit	Hours	Hours £	Mat'ls £	O & P £	Total £
exceeding 300mm girth	m2	0.78	9.36	2.23	1.74	13.33

One coat primer, two coats
undercoat and one coat gloss on
metal radiators

	Unit	Hours	Hours £	Mat'ls £	O & P £	Total £
exceeding 300mm girth	m2	0.72	8.64	2.19	1.62	12.45

On external woodwork

One coat primer on wood
general surfaces

	Unit	Hours	Hours £	Mat'ls £	O & P £	Total £
exceeding 300mm girth	m2	0.24	2.88	0.54	0.51	3.93
not exceeding 150mm girth	m	0.06	0.72	0.09	0.12	0.93
150-300mm girth	m	0.08	0.96	0.17	0.17	1.30
isolated areas not exceeding 0.5m2	nr	0.02	0.24	0.26	0.08	0.58

One coat primer on wood
windows, glazed doors and the like

	Unit	Hours	Hours £	Mat'ls £	O & P £	Total £
panes area not exceeding 0.1m2	m2	0.40	4.80	0.19	0.75	5.74
panes area 0.1 to 0.5m2	m2	0.35	4.20	0.17	0.66	5.03
panes area 0.5 to 1m2	m2	0.30	3.60	0.15	0.56	4.31
panes area exceeding 1m2	m2	0.25	3.00	0.13	0.47	3.60

	Unit	Hours	Hours £	Mat'ls £	O & P £	Total £
One coat primer on wood frames and the like						
exceeding 300mm girth	m2	0.26	3.12	0.50	0.54	4.16
not exceeding 150mm girth	m	0.07	0.84	0.07	0.14	1.05
150-300mm girth	m	0.10	1.20	0.15	0.20	1.55
isolated areas not exceeding 0.5m2	nr	0.14	1.68	0.23	0.29	2.20
One coat undercoat on primed wood general surfaces						
exceeding 300mm girth	m2	0.24	2.88	0.54	0.51	3.93
not exceeding 150mm girth	m	0.06	0.72	0.09	0.12	0.93
150-300mm girth	m	0.08	0.96	0.17	0.17	1.30
isolated areas not exceeding 0.5m2	nr	0.12	1.44	0.26	0.26	1.96
One coat undercoat on primed wood windows, glazed doors and the like						
panes area not exceeding 0.1m2	m2	0.40	4.80	0.21	0.75	5.76
panes area 0.1 to 0.5m2	m2	0.35	4.20	0.19	0.66	5.05
panes area 0.5 to 1m2	m2	0.30	3.60	0.17	0.57	4.34
panes area exceeding 1m2	m2	0.25	3.00	0.15	0.47	3.62
One coat undercoat on primed wood frames and the like						
exceeding 300mm girth	m2	0.26	3.12	0.54	0.55	4.21
not exceeding 150mm girth	m	0.07	0.84	0.09	0.14	1.07
150-300mm girth	m	0.10	1.20	0.17	0.21	1.58
isolated areas not exceeding 0.5m2	nr	0.14	1.68	0.26	0.29	2.23

	Unit	Hours	Hours £	Mat'ls £	O & P £	Total £
Two coats undercoat on primed wood general surfaces						
exceeding 300mm girth	m2	0.48	5.76	1.09	1.03	7.88
not exceeding 150mm girth	m	0.12	1.44	0.17	0.24	1.85
150-300mm girth	m	0.16	1.92	0.34	0.34	2.60
isolated areas not exceeding 0.5m2	nr	0.24	2.88	0.52	0.51	3.91
Two coats undercoat on primed wood windows, glazed doors and the like						
panes area not exceeding 0.1m2	m2	0.80	9.60	0.43	1.50	11.53
panes area 0.1 to 0.5m2	m2	0.70	8.40	0.39	1.32	10.11
panes area 0.5 to 1m2	m2	0.60	7.20	0.34	1.13	8.67
panes area exceeding 1m2	m2	0.50	6.00	0.30	0.95	7.25
Two coats undercoat on primed wood frames and the like						
exceeding 300mm girth	m2	0.52	6.24	1.09	1.10	8.43
not exceeding 150mm girth	m	0.14	1.68	0.17	0.28	2.13
150-300mm girth	m	0.20	2.40	0.34	0.41	3.15
isolated areas not exceeding 0.5m2	nr	0.32	3.84	0.52	0.65	5.01
One coat gloss on undercoated wood general surfaces						
exceeding 300mm girth	m2	0.24	2.88	0.54	0.51	3.93
not exceeding 150mm girth	m	0.06	0.72	0.09	0.12	0.93
150-300mm girth	m	0.08	0.96	0.17	0.17	1.30
isolated areas not exceeding 0.5m2	nr	0.12	1.44	0.26	0.26	1.96

	Unit	Hours	Hours £	Mat'ls £	O & P £	Total £
One coat gloss on undercoated wood windows, glazed doors and the like						
panes area not exceeding 0.1m2	m2	0.40	4.80	0.21	0.75	5.76
panes area 0.1 to 0.5m2	m2	0.35	4.20	0.09	0.64	4.93
panes area 0.5 to 1m2	m2	0.30	3.60	0.17	0.57	4.34
panes area exceeding 1m2	m2	0.25	3.00	0.15	0.47	3.62
One coat gloss on undercoated wood frames and the like						
exceeding 300mm girth	m2	0.26	3.12	0.54	0.55	4.21
not exceeding 150mm girth	m	0.07	0.84	0.09	0.14	1.07
150-300mm girth	m	0.10	1.20	0.17	0.21	1.58
isolated areas not exceeding 0.5m2	nr	0.16	1.92	0.26	0.33	2.51
One coat primer, one coat undercoat and one coat gloss on wood general surfaces						
exceeding 300mm girth	m2	0.73	8.70	1.63	1.55	11.88
not exceeding 150mm girth	m	0.14	1.68	0.26	0.29	2.23
150-300mm girth	m	0.24	2.88	0.52	0.51	3.91
isolated areas not exceeding 0.5m2	nr	0.40	4.80	0.78	0.84	6.42
One coat primer, one coat undercoat and one coat gloss on wood windows, glazed doors and the like						
panes area not exceeding 0.1m2	m2	1.20	14.40	0.63	2.25	17.28
panes area 0.1 to 0.5m2	m2	1.05	12.60	0.47	1.96	15.03
panes area 0.5 to 1m2	m2	0.90	10.80	0.50	1.70	13.00
panes area exceeding 1m2	m2	0.75	9.00	0.43	1.41	10.84

	Unit	Hours	Hours £	Mat'ls £	O & P £	Total £
One coat primer, one coat undercoat and one coat gloss on wood frames and the like						
exceeding 300mm girth	m2	0.78	9.36	1.59	1.64	12.59
not exceeding 150mm girth	m	0.21	2.52	0.23	0.41	3.16
150-300mm girth	m	0.30	3.60	0.50	0.62	4.72
isolated areas not exceeding 0.5m2	nr	0.48	5.76	0.76	0.98	7.50
One coat primer, two coats undercoat and one coat gloss on wood general surfaces						
exceeding 300mm girth	m2	0.96	11.52	2.18	2.06	15.76
not exceeding 150mm girth	m	0.25	3.00	0.34	0.50	3.84
150-300mm girth	m	0.34	4.08	0.69	0.72	5.49
isolated areas not exceeding 0.5m2	nr	0.48	5.76	1.04	1.02	7.82
One coat primer, two coats undercoat and one coat gloss on wood windows, glazed doors and the like						
panes area not exceeding 0.1m2	m2	1.60	19.20	0.85	3.01	23.06
panes area 0.1 to 0.5m2	m2	1.40	16.80	0.76	2.63	20.19
panes area 0.5 to 1m2	m2	1.20	14.40	0.67	2.26	17.33
panes area exceeding 1m2	m2	1.00	12.00	0.58	1.89	14.47
One coat primer, two coats undercoat and one coat gloss on wood frames and the like						
exceeding 300mm girth	m2	1.04	12.48	2.13	2.19	16.80
not exceeding 150mm girth	m	0.28	3.36	0.32	0.55	4.23
150-300mm girth	m	0.40	4.80	0.67	0.82	6.29
isolated areas not exceeding .5m2	nr	0.64	7.68	1.02	1.31	10.01

	Unit	Hours	Hours £	Mat'ls £	O & P £	Total £
On external metalwork						
One coat primer on metal general surfaces						
exceeding 300mm girth	m2	0.24	2.88	0.59	0.52	3.99
not exceeding 150mm girth	m	0.06	0.72	0.11	0.12	0.95
150-300mm girth	m	0.08	0.96	0.19	0.17	1.32
isolated areas not exceeding 0.5m2	nr	0.12	1.44	0.30	0.26	2.00
One coat primer on metal windows, glazed doors and the like						
panes area not exceeding 0.1m2	m2	0.40	4.80	0.21	0.75	5.76
panes area 0.1 to 0.5m2	m2	0.35	4.20	0.19	0.66	5.05
panes area 0.5 to 1m2	m2	0.30	3.60	0.17	0.57	4.34
panes area exceeding 1m2	m2	0.25	3.00	0.15	0.47	3.62
One coat primer on structural steelwork						
exceeding 300mm girth	m2	0.30	3.60	0.59	0.63	4.82
One coat primer on pipes and gutters						
exceeding 300mm girth	m2	0.26	3.12	0.59	0.56	4.27
150-300mm girth	m	0.08	0.96	0.19	0.17	1.32
One coat undercoat on metal general surfaces						
exceeding 300mm girth	m2	0.24	2.88	0.54	0.51	3.93
not exceeding 150mm girth	m	0.06	0.72	0.09	0.12	0.93
150-300mm girth	m	0.08	0.96	0.17	0.17	1.30
isolated areas not exceeding 0.5m2	nr	0.12	1.44	0.26	0.26	1.96

	Unit	Hours	Hours £	Mat'ls £	O & P £	Total £
One coat undercoat on metal windows, glazed doors and the like						
panes area not exceeding 0.1m2	m2	0.40	4.80	0.19	0.75	5.74
panes area 0.1 to 0.5m2	m2	0.35	4.20	0.17	0.66	5.03
panes area 0.5 to 1m2	m2		0.00	0.15	0.02	0.17
panes area exceeding 1m2	m2	0.25	3.00	0.13	0.47	3.60
One coat undercoat on structural steelwork						
exceeding 300mm girth	m2	0.30	3.60	0.54	0.62	4.76
One coat primer on pipes and gutters						
exceeding 300mm girth	m2	0.26	3.12	0.54	0.55	4.21
150-300mm girth	m	0.08	0.96	0.17	0.17	1.30
Two coats undercoat on metal general surfaces						
exceeding 300mm girth	m2	0.48	5.76	1.09	1.03	7.88
not exceeding 150mm girth	m	0.12	1.44	0.17	0.24	1.85
150-300mm girth	m	0.20	2.40	0.34	0.41	3.15
isolated areas not exceeding 0.5m2	nr	0.24	2.88	0.52	0.51	3.91
Two coats undercoat on metal windows, glazed doors and the like						
panes area not exceeding 0.1m2	m2	0.80	9.60	0.39	1.50	11.49
panes area 0.1 to 0.5m2	m2	0.70	8.40	0.34	1.31	10.05
panes area 0.5 to 1m2	m2	0.60	7.20	0.30	1.13	8.63
panes area exceeding 1m2	m2	0.50	6.00	0.26	0.94	7.20

	Unit	Hours	Hours £	Mat'ls £	O & P £	Total £
Two coats undercoat on structural steelwork						
exceeding 300mm girth	m2	0.60	7.20	1.09	1.24	9.53
Two coats undercoat on pipes and gutters						
exceeding 300mm girth	m2	0.52	6.24	1.09	1.10	8.43
150-300mm girth	m	0.16	1.92	0.34	0.34	2.60
One coat gloss on metal general surfaces						
exceeding 300mm girth	m2	0.24	2.88	0.54	0.51	3.93
not exceeding 150mm girth	m	0.06	0.72	0.09	0.12	0.93
150-300mm girth	m	0.08	0.96	0.17	0.17	1.30
isolated areas not exceeding 0.5m2	nr	0.12	1.44	0.26	0.26	1.96
One coat gloss on metal windows, glazed doors and the like						
panes area not exceeding 0.1m2	m2	0.40	4.80	0.19	0.75	5.74
panes area 0.1 to 0.5m2	m2	0.35	4.20	0.17	0.66	5.03
panes area 0.5 to 1m2	m2	0.30	3.60	0.15	0.56	4.31
panes area exceeding 1m2	m2	0.25	3.00	0.13	0.47	3.60
One coat gloss on structural steelwork						
exceeding 300mm girth	m2	0.30	3.60	0.54	0.62	4.76
One coat gloss on pipes and gutters						
exceeding 300mm girth	m2	0.26	3.12	0.54	0.55	4.21
150-300mm girth	m	0.08	0.96	0.19	0.17	1.32

	Unit	Hours	Hours £	Mat'ls £	O & P £	Total £
One coat primer, one coat undercoat and one coat gloss on metal general surfaces						
exceeding 300mm girth	m2	0.72	8.64	1.68	1.55	11.87
not exceeding 150mm girth	m	0.18	2.16	0.28	0.37	2.81
150-300mm girth	m	0.24	2.88	0.54	0.51	3.93
isolated areas not exceeding 0.5m2	nr	0.36	4.32	0.82	0.77	5.91
One coat primer, one coat undercoat and one coat gloss on metal windows, glazed doors and the like						
panes area not exceeding 0.1m2	m2	1.20	14.40	0.65	2.26	17.31
panes area 0.1 to 0.5m2	m2	1.05	12.60	0.50	1.97	15.07
panes area 0.5 to 1m2	m2	0.90	10.80	0.52	1.70	13.02
panes area exceeding 1m2	m2	0.75	9.00	0.45	1.42	10.87
One coat primer, one coat undercoat and one coat gloss on structural steelwork						
exceeding 300mm girth	m2	0.90	10.80	1.68	1.87	14.35
One coat primer, one coat undercoat and one coat gloss on pipes and gutters						
exceeding 300mm girth	m2	0.78	9.36	1.63	1.65	12.64
150-300mm girth	m	0.24	2.88	0.58	0.52	3.98

	Unit	Hours	Hours £	Mat'ls £	O & P £	Total £
One coat primer, two coats undercoat and one coat gloss on metal general surfaces						
exceeding 300mm girth	m2	0.96	11.52	2.23	2.06	15.81
not exceeding 150mm girth	m	0.24	2.88	0.28	0.47	3.63
150-300mm girth	m	0.32	3.84	0.54	0.66	5.04
isolated areas not exceeding 0.5m2	nr	0.48	5.76	0.80	0.98	7.54
One coat primer, two coats undercoat and one coat gloss on metal windows, glazed doors and the like						
panes area not exceeding 0.1m2	m2	1.60	19.20	0.80	3.00	23.00
panes area 0.1 to 0.5m2	m2	1.40	16.80	0.71	2.63	20.14
panes area 0.5 to 1m2	m2	1.20	14.40	0.65	2.26	17.31
panes area exceeding 1m2	m2	1.00	12.00	0.54	1.88	14.42
One coat primer, two coats undercoat and one coat gloss on structural steelwork						
exceeding 300mm girth	m2	1.20	14.40	2.23	2.49	19.12
One coat primer, two coats undercoat and one coat gloss on pipes and gutters						
exceeding 300mm girth	m2	1.04	12.48	2.18	2.20	16.86
150-300mm girth	m	0.32	3.84	0.82	0.70	5.36

	Unit	Hours	Hours £	Mat'ls £	O & P £	Total £

Wallpapering

Prepare, size, apply adhesive, supply
and hang paper to plastered walls
and columns, butt jointed

	Unit	Hours	Hours £	Mat'ls £	O & P £	Total £
lining paper						
£1.50 per roll	m2	0.30	3.60	0.36	0.59	4.55
£2.00 per roll	m2	0.30	3.60	0.48	0.61	4.69
£2.50 per roll	m2	0.30	3.60	0.59	0.63	4.82
washable paper						
£2.50 per roll	m2	0.30	3.60	0.59	0.63	4.82
£4.00 per roll	m2	0.30	3.60	0.94	0.68	5.22
£5.00 per roll	m2	0.30	3.60	1.18	0.72	5.50
vinyl paper						
£4.00 per roll	m2	0.30	3.60	0.94	0.68	5.22
£5.00 per roll	m2	0.30	3.60	1.18	0.72	5.50
£6.00 per roll	m2	0.30	3.60	1.41	0.75	5.76
washable paper						
£5.00 per roll	m2	0.30	3.60	1.18	0.72	5.50
£6.00 per roll	m2	0.30	3.60	1.41	0.75	5.76
£7.00 per roll	m2	0.30	3.60	1.64	0.79	6.03
hessian paper						
£7.00 per m2	m2	0.50	6.00	7.70	2.06	15.76
£8.00 per m2	m2	0.50	6.00	8.80	2.22	17.02
£9.00 per m2	m2	0.50	6.00	9.90	2.39	18.29
suede paper						
£9.00 per m2	m2	0.30	3.60	9.90	2.03	15.53
£10.00 per m2	m2	0.30	3.60	11.00	2.19	16.79
£11.00 per m2	m2	0.30	3.60	12.10	2.36	18.06

	Unit	Hours	Hours £	Mat'ls £	O & P £	Total £

Prepare, size, apply adhesive, supply
and hang paper to plastered ceilings
and columns, butt jointed

	Unit	Hours	Hours £	Mat'ls £	O & P £	Total £
lining paper						
£1.50 per roll	m2	0.35	4.20	0.36	0.68	5.24
£2.00 per roll	m2	0.35	4.20	0.48	0.70	5.38
£2.50 per roll	m2	0.35	4.20	0.59	0.72	5.51
washable paper						
£2.50 per roll	m2	0.35	4.20	0.59	0.72	5.51
£4.00 per roll	m2	0.35	4.20	0.94	0.77	5.91
£5.00 per roll	m2	0.35	4.20	1.18	0.81	6.19
vinyl paper						
£4.00 per roll	m2	0.35	4.20	0.94	0.77	5.91
£5.00 per roll	m2	0.35	4.20	1.18	0.81	6.19
£6.00 per roll	m2	0.35	4.20	1.41	0.84	6.45
washable paper						
£5.00 per roll	m2	0.35	4.20	1.18	0.81	6.19
£6.00 per roll	m2	0.35	4.20	1.41	0.84	6.45
£7.00 per roll	m2	0.35	4.20	1.64	0.88	6.72

	Unit	Hours	Hours £	Mat'ls £	O & P £	Total £

EXTERNAL WORKS

Drainage

Excavate drain trench by hand,
support sides, level and ram
bottom of trench, backfill and
consolidate with excavated
material and remove surplus to
skip for pipes 100mm and
150mm diameter, average depth
of trench

	Unit	Hours	Hours £	Mat'ls £	O & P £	Total £
0.50m	m	1.20	11.76	0.00	1.76	13.52
0.75m	m	1.90	18.62	0.00	2.79	21.41
1.00m	m	2.50	24.50	0.00	3.68	28.18
1.25m	m	4.00	39.20	0.00	5.88	45.08
1.50m	m	5.00	49.00	0.00	7.35	56.35
1.75m	m	6.00	58.80	0.00	8.82	67.62
2.00m	m	7.20	70.56	0.00	10.58	81.14
2.25m	m	8.20	80.36	0.00	12.05	92.41
2.50m	m	9.50	93.10	0.00	13.97	107.07
2.75m	m	10.50	102.90	0.00	15.44	118.34
3.00m	m	12.00	117.60	0.00	17.64	135.24

Excavate drain trench by hand,
support sides, level and ram
bottom of trench, backfill and
consolidate with excavated
material and remove surplus to
skip for pipes 225mm diameter,
average depth of trench

	Unit	Hours	Hours £	Mat'ls £	O & P £	Total £
0.50m	m	1.30	12.74	0.00	1.91	14.65
0.75m	m	2.10	20.58	0.00	3.09	23.67
1.00m	m	2.70	26.46	0.00	3.97	30.43
1.25m	m	4.30	42.14	0.00	6.32	48.46
1.50m	m	5.40	52.92	0.00	7.94	60.86
1.75m	m	6.60	64.68	0.00	9.70	74.38

	Unit	Hours	Hours £	Mat'ls £	O & P £	Total £
2.00m	m	7.10	69.58	0.00	10.44	80.02
2.25m	m	9.00	88.20	0.00	13.23	101.43
2.50m	m	10.20	99.96	0.00	14.99	114.95
2.75m	m	11.30	110.74	0.00	16.61	127.35
3.00m	m	12.80	125.44	0.00	18.82	144.26

Extra for breaking up by hand

	Unit	Hours	Hours £	Mat'ls £	O & P £	Total £
plain concrete 100mm thick	m2	0.90	8.82	0.00	1.32	10.14
reinforced concrete 100mm thick	m2	1.00	9.80	0.00	1.47	11.27
tarmacadam 75mm thick	m2	0.55	5.39	0.00	0.81	6.20
hardcore 100mm thick	m2	0.40	3.92	0.00	0.59	4.51
soft rock	m3	8.00	78.40	0.00	11.76	90.16
hard rock	m3	10.00	98.00	0.00	14.70	112.70

	Unit	Hours	Hours £	Plant £	O & P £	Total £

Excavate drain trench by machine, support sides, level and ram bottom of trench, backfill and consolidate with excavated material and remove surplus to skip for pipes 100mm and 150mm diameter, average depth of trench

	Unit	Hours	Hours £	Plant £	O & P £	Total £
0.50m	m	0.30	3.24	1.99	0.78	6.01
0.75m	m	0.50	5.40	2.99	1.26	9.65
1.00m	m	0.90	9.72	4.44	2.12	16.28
1.25m	m	1.50	16.20	6.10	3.35	25.65
1.50m	m	1.80	19.44	6.66	3.92	30.02
1.75m	m	2.20	23.76	7.77	4.73	36.26
2.00m	m	2.80	30.24	8.88	5.87	44.99
2.25m	m	3.40	36.72	11.10	7.17	54.99
2.50m	m	3.90	42.12	12.76	8.23	63.11
2.75m	m	4.50	48.60	14.43	9.45	72.48
3.00m	m	5.00	54.00	16.09	10.51	80.60

	Unit	Hours	Hours £	Plant £	O & P £	Total £

Excavate drain trench by machine,
support sides, level and ram
bottom of trench, backfill and
consolidate with excavated
material and remove surplus to
skip for pipes 225mm diameter,
average depth of trench

	Unit	Hours	Hours £	Plant £	O & P £	Total £
0.50m	m	1.30	14.04	1.99	2.40	18.43
0.75m	m	2.10	22.68	2.99	3.85	29.52
1.0m	m	2.70	29.16	4.55	5.06	38.77
1.25m	m	4.30	46.44	6.38	7.92	60.74
1.50m	m	5.40	58.32	6.93	9.79	75.04
1.75m	m	6.60	71.28	8.21	11.92	91.41
2.00m	m	7.10	76.68	9.49	12.93	99.10
2.25m	m	9.00	97.20	11.32	16.28	124.80
2.50m	m	10.20	110.16	12.76	18.44	141.36
2.75m	m	11.30	122.04	14.98	20.55	157.57
3.00m	m	12.80	138.24	16.37	23.19	177.80

Extra for breaking up by machine

	Unit	Hours	Hours £	Plant £	O & P £	Total £
plain concrete 100mm thick	m2	0.30	3.24	1.66	0.74	5.64
reinforced concrete 100mm thick	m2	0.35	3.78	1.99	0.87	6.64
tarmacadam 75mm thick	m2	0.20	2.16	1.22	0.51	3.89
hardcore 100mm thick	m2	0.15	1.62	0.88	0.38	2.88
soft rock	m3	4.55	49.14	6.43	8.34	63.91
hard rock	m3	5.20	56.16	8.04	9.63	73.83

Sand bed in trench under pipe
100mm diameter, thickness

	Unit	Hours	Hours £	Plant £	O & P £	Total £
100mm	m	0.10	0.98	1.05	0.30	2.33
150mm	m	0.12	1.18	1.55	0.41	3.13

	Unit	Hours	Hours £	Mat'ls £	O & P £	Total £
Sand bed in trench under pipe 150mm diameter, thickness						
100mm	m	0.11	1.08	1.44	0.38	2.90
150mm	m	0.13	1.27	1.60	0.43	3.31
Sand bed in trench under pipe 225mm diameter, thickness						
100mm	m	0.14	1.37	1.72	0.46	3.56
150mm	m	0.16	1.57	2.49	0.61	4.67
Granular filling in bed in trench under pipe 100mm diameter, thickness						
100mm	m	0.12	1.18	0.88	0.31	2.36
150mm	m	0.14	1.37	1.28	0.40	3.05
Granular filling in bed in trench under pipe 150mm diameter, thickness						
100mm	m	0.13	1.27	0.99	0.34	2.60
150mm	m	0.15	1.47	1.49	0.44	3.40
Granular filling in bed in trench under pipe 225mm diameter, thickness						
100mm	m	0.16	1.57	1.27	0.43	3.26
150mm	m	0.18	1.76	1.94	0.56	4.26
Concrete in bed in trench under pipe, 100mm diameter, thickness						
100mm	m	0.24	2.35	3.88	0.93	7.17
150mm	m	0.28	2.74	5.77	1.28	9.79

	Unit	Hours	Hours £	Mat'ls £	O & P £	Total £
Concrete in bed in trench under pipe, 150mm diameter, thickness						
100mm	m	0.26	2.55	4.49	1.06	8.09
150mm	m	0.30	2.94	5.71	1.30	9.95
Concrete in bed in trench under pipe, 225mm diameter, thickness						
100mm	m	0.32	3.14	5.10	1.24	9.47
150mm	m	0.36	3.53	7.82	1.70	13.05
Concrete in bed and haunching to pipe, 100mm diameter, bed thickness						
100mm	m	0.48	4.70	7.10	1.77	13.57
150mm	m	0.56	5.49	10.61	2.41	18.51
Concrete in bed and haunching to pipe, 150mm diameter, bed thickness						
100mm	m	0.52	5.10	8.02	1.97	15.08
150mm	m	0.60	5.88	10.96	2.53	19.37
Concrete in bed and haunching to pipe, 225mm diameter, bed thickness						
100mm	m	0.60	5.88	9.03	2.24	17.15
150mm	m	0.66	6.47	11.69	2.72	20.88
Granular filling in bed and surround to pipe, 100mm diameter, thickness						
100mm	m	0.36	3.53	3.56	1.06	8.15
150mm	m	0.42	4.12	5.50	1.44	11.06

	Unit	Hours	Hours £	Mat'ls £	O & P £	Total £
Granular filling in bed and surround to pipe, 150mm diameter, thickness						
100mm	m	0.40	3.92	5.02	1.34	10.28
150mm	m	0.46	4.51	7.60	1.82	13.92
Granular filling in bed and surround to pipe, 225mm diameter, thickness						
100mm	m	0.44	4.31	7.54	1.78	13.63
150mm	m	0.50	4.90	9.01	2.09	16.00
Concrete in bed and surround to pipe, 100mm diameter, thickness						
100mm	m	0.72	7.06	8.43	2.32	17.81
150mm	m	0.84	8.23	9.17	2.61	20.01
Concrete in bed and surround to pipe, 150mm diameter, thickness						
100mm	m	0.72	7.06	9.16	2.43	18.65
150mm	m	0.84	8.23	10.01	2.74	20.98
Concrete in bed and surround to pipe, 225mm diameter, thickness						
100mm	m	0.72	7.06	11.85	2.84	21.74
150mm	m	0.84	8.23	13.96	3.33	25.52

	Unit	Hours	Hours £	Mat'ls £	O & P £	Total £
Hepworths' Supersleve vitrified clay drain pipes, spigot and socket joints with sealing rings, 100mm diameter						
laid in trenches	m	0.36	3.89	6.39	1.54	11.82
in lengths not exceeding 3m	m	0.54	5.83	6.39	1.83	14.06
bend	nr	0.30	3.24	8.70	1.79	13.73
rest bend	nr	0.30	3.24	13.53	2.52	19.29
single junction	nr	0.30	3.24	12.76	2.40	18.40
Hepworths' Supersleve vitrified clay drain pipes, spigot and socket joints with sealing rings, 150mm diameter						
laid in trenches	m	0.60	6.48	12.19	2.80	21.47
in lengths not exceeding 3m	m	0.40	4.32	12.19	2.48	18.99
bend	nr	0.30	3.24	12.16	2.31	17.71
rest bend	nr	0.30	3.24	15.63	2.83	21.70
single junction	nr	0.30	3.24	16.29	2.93	22.46
Vitrified clay inlet gulley complete with grid and surrounded in concrete	nr	1.50	16.20	40.58	8.52	65.30
Vitrified clay back inlet gulley complete with grid and surrounded in concrete	nr	1.50	16.20	57.99	11.13	85.32
Vitrified clay paved area gulley complete with grid and surrounded in concrete	nr	1.50	16.20	47.17	9.51	72.88

	Unit	Hours	Hours £	Mat'ls £	O & P £	Total £

Manholes

Excavate by hand for manhole
not exceeding

1.0m deep	m3	4.00	39.20	0.00	5.88	45.08
1.5m deep	m3	4.50	44.10	0.00	6.62	50.72
2.0m deep	m3	5.60	54.88	0.00	8.23	63.11

Dispose surplus excavated material
by hand, load into barrows, wheel

50m and deposit into skip	m3	2.20	21.56	0.00	3.23	24.79

Excavate by machine for manhole
not exceeding

1.0m deep	m3	0.35	3.43	4.66	1.21	9.30
1.5m deep	m3	0.40	3.92	4.88	1.32	10.12
2.0m deep	m3	0.45	4.41	5.10	1.43	10.94

Earthwork support not exceeding
2m between opposing faces,
depth not exceeding

1.0m deep	m2	0.35	3.43	1.44	0.73	5.60
2.0m deep	m2	0.40	3.92	1.55	0.82	6.29
4.0m deep	m2	0.45	4.41	1.66	0.91	6.98

Site-mixed concrete in base of
manhole, thickness

100-150mm	m3	2.00	19.60	68.98	13.29	101.87
150-300mm	m3	1.80	17.64	68.98	12.99	99.61

Site-mixed concrete in benching
to manhole, average thickness

225mm	m3	6.00	58.80	77.11	20.39	156.30

	Unit	Hours	Hours £	Mat'ls £	O & P £	Total £
Engineering bricks Class 'B' in cement mortar in walls of manhole	m2	4.80	57.60	55.38	16.95	129.93
Extra for fair face and flush pointing	m2	0.25	3.00	0.00	0.45	3.45
Build in ends of pipes to one brick wall and make good, pipe diameter						
100mm	nr	0.15	1.80	0.00	0.27	2.07
150mm	nr	0.19	2.27	0.00	0.34	2.61
Galvanised step iron built into brickwork	nr	0.20	2.40	5.96	1.25	9.61
Cast iron manhole cover, frame bedded in cement mortar						
Grade A, light duty, size 600 × 450mm	nr	1.90	22.80	65.46	13.24	101.50
Grade B, medium duty, size 600 × 450mm	nr	2.00	24.00	113.63	20.64	158.27
Best quality vitrified clay channels, bedded in cement mortar						
half-section straight main channel						
100mm diameter × 300mm	nr	0.75	8.10	2.94	1.66	12.70
100mm diameter × 600mm	nr	0.85	9.18	2.94	1.82	13.94
100mm diameter × 1000mn	nr	0.95	10.26	4.11	2.16	16.53
150mm diameter × 300mm	nr	0.75	8.10	4.95	1.96	15.01
150mm diameter × 1000mn	nr	0.85	9.18	4.95	2.12	16.25
150mm diameter × 1000mn	nr	0.95	10.26	7.24	2.63	20.13
Half-section 90 degrees channel bend						
100mm	nr	0.30	3.24	4.17	1.11	8.52
150mm	nr	0.40	4.32	7.20	1.73	13.25

	Unit	Hours	Hours £	Mat'ls £	O & P £	Total £
Three-quarter section 90 degrees channel bend						
100mm	nr	1.20	12.96	11.76	3.71	28.43
150mm	nr	1.30	14.04	20.46	5.18	39.68

	Unit	Spec'list price £	O & P £	Total £
Fencing				
Chainlink fencing, galvanised steel mesh on three strained line wires fixed to concrete posts at 3000mm centres				
900mm high	m	14.20	2.13	16.33
1200mm high	m	17.38	2.61	19.99
1400mm high	m	19.54	2.93	22.47
1800mm high	m	26.05	3.91	29.96
Chainlink fencing, galvanised steel mesh on three strained line wires fixed to galvanised mild steel posts at 3000mm centres				
900mm high	m	15.26	2.29	17.55
1200mm high	m	18.05	2.71	20.76
1400mm high	m	21.31	3.20	24.51
1800mm high	m	27.72	4.16	31.88
Chainlink fencing, plastic-coated mild steel mesh on three strained line wires fixed to concrete posts at 3000mm centres				
900mm high	m	14.81	2.22	17.03
1200mm high	m	18.21	2.73	20.94
1400mm high	m	20.37	3.06	23.43
1800mm high	m	26.88	4.03	30.91

	Unit	Spec'list price £	O & P £	Total £
Chainlink fencing, plastic-coated mild steel mesh on three strained line wires fixed to galvanised mild steel posts at 3000mm centres				
900mm high	m	16.09	2.41	18.50
1200mm high	m	18.89	2.83	21.72
1400mm high	m	23.14	3.47	26.61
1800mm high	m	28.56	4.28	32.84
Strained wire fence including concrete posts at 3000mm centres				
1.00m high, 5 wires	m	12.07	1.81	13.88
1.20m high, 6 wires	m	12.40	1.86	14.26
1.40m high, 8 wires	m	13.07	1.96	15.03
Cleft chestnut fencing, pales set 75mm apart, 75mm diameter softwood posts at 2500mm centres				
0.90m high, 2 wires	m	10.87	1.63	12.50
1.05m high, 2 wires	m	11.15	1.67	12.82
1.20m high, 2 wires	m	11.72	1.76	13.48
1.50m high, 3 wires	m	12.80	1.92	14.72
1.80m high, 4 wires	m	13.84	2.08	15.92
Close-boarded fencing consisting of 90 x 19mm pales lapped 13mm fixed to 75 x 38mm horizontal rails on 75 x 75mm posts at 3000mm centres, height				
1.00m	m	34.68	5.20	39.88
1.20m	m	36.35	5.45	41.80
1.40m	m	38.01	5.70	43.71
1.60m	m	39.68	5.95	45.63
1.80m	m	41.34	6.20	47.54

	Unit	Hours	Hours £	Mat'ls £	O & P £	Total £

Kerbs and edgings

Excavate trench by hand for kerb
foundation

200 × 75mm	m	0.10	1.08	0.00	0.16	1.24
250 × 100mm	m	0.12	1.30	0.00	0.19	1.49
300 × 100mm	m	0.14	1.51	0.00	0.23	1.74
450 × 150mm	m	0.20	2.16	0.00	0.32	2.48
600 × 200mm	m	0.40	4.32	0.00	0.65	4.97

Excavate curved trench by hand
for kerb foundation

250 × 100mm	m	0.12	1.30	0.00	0.19	1.49
250 × 100mm	m	0.14	1.51	0.00	0.23	1.74
300 × 100mm	m	0.16	1.73	0.00	0.26	1.99
600 × 200mm	m	0.24	2.59	0.00	0.39	2.98
600 × 200mm	m	0.44	4.75	0.00	0.71	5.46

	Unit	Hours	Hours £	Plant £	O & P £	Total £

Excavate trench by machine for
kerb foundation
200 × 75mm

200 × 75mm	m	0.00	0.00	0.72	0.11	0.83
250 × 100mm	m	0.00	0.00	0.77	0.12	0.89
600 200mm	m	0.00	0.00	0.83	0.12	0.95
	m	0.00	0.00	1.36	0.20	1.56
	m	0.00	0.00	2.44	0.37	2.81

Excavate curved trench by machine
for kerb foundation

200 × 75mm	m	0.00	0.00	0.83	0.12	0.95
250 × 100mm	m	0.00	0.00	0.88	0.13	1.01
300 × 100mm	m	0.00	0.00	0.94	0.14	1.08
450 × 150mm	m	0.00	0.00	1.62	0.24	1.86
600 × 200mm	m	0.00	0.00	2.85	0.43	3.28

	Unit	Hours	Hours £	Mat'ls £	O & P £	Total £
Site-mixed concrete in foundation for kerb size						
200 × 75mm	m	0.03	0.29	1.04	0.20	1.53
250 × 100mm	m	0.04	0.39	1.72	0.32	2.43
300 × 100mm	m	0.05	0.49	2.06	0.38	2.93
450 × 150mm	m	0.18	1.76	4.66	0.96	7.39
600 × 200mm	m	0.30	2.94	8.26	1.68	12.88
Precast concrete kerbs, channels and edgings, jointed and pointed in cement mortar						
kerbs, straight						
127 × 254mm	m	0.40	3.92	5.22	1.37	10.51
152 × 305mm	m	0.45	4.41	7.63	1.81	13.85
kerbs, curved						
127 × 254mm	m	0.50	4.90	7.38	1.84	14.12
152 × 305mm	m	0.55	5.39	9.40	2.22	17.01
channels, straight						
127 × 254mm	m	0.40	3.92	5.52	1.42	10.86
channels, curved						
127 × 254mm	m	0.50	4.90	7.38	1.84	14.12
edgings, straight						
51 × 152mm	m	0.30	2.94	3.15	0.91	7.00
51 × 203mm	m	0.30	2.94	3.68	0.99	7.61

Sub-bases

Beds and bases compacting in layers and grading

	Unit	Hours	Hours £	Mat'ls £	O & P £	Total £
average thickness, 100mm						
granular fill	m2	0.10	0.98	2.88	0.58	4.44
sand	m2	0.12	1.18	2.66	0.58	4.41
hardcore	m2	0.13	1.27	2.55	0.57	4.40

	Unit	Hours	Hours £	Mat'ls £	O & P £	Total £
average thickness, 150mm						
granular fill	m2	0.12	1.18	3.59	0.71	5.48
sand	m2	0.14	1.37	3.44	0.72	5.53
hardcore	m2	0.16	1.57	3.27	0.73	5.56
average thickness, 200mm						
granular fill	m2	0.14	1.37	4.66	0.90	6.94
sand	m2	0.16	1.57	4.21	0.87	6.64
hardcore	m2	0.18	1.76	4.05	0.87	6.69

Concrete beds

Site-mixed concrete in beds

not exceeding 150mm thick	m3	2.80	27.44	77.11	15.68	120.23
150 to 450mm thick	m3	2.20	21.56	77.11	14.80	113.47

Formwork to sides of concrete
bases

not exceeding 250mm wide	m	0.65	6.37	1.79	1.22	9.38
250 to 500mm wide	m	0.90	8.82	2.93	1.76	13.51

Steel fabric reinforcement laid in
concrete beds

ref A142, 2.22kg/m2	m2	0.15	1.47	1.37	0.43	3.27
ref A193, 3.02kg/m2	m2	0.18	1.76	1.65	0.51	3.93

Expansion joint, impregnated
fibre-base joint filler, formed joint

12.5mm thick						
not exceeding 150mm wid	m	0.18	1.76	2.33	0.61	4.71
150 to 300mm wide	m	0.24	2.35	3.58	0.89	6.82
300 to 450mm wide	m	0.30	2.94	5.40	1.25	9.59

	Unit	Hours	Hours £	Mat'ls £	O & P £	Total £
25mm thick						
not exceeding 150mm wid	m	0.20	1.96	3.18	0.77	5.91
150 to 300mm wide	m	0.26	2.55	4.73	1.09	8.37
300 to 450mm wide	m	0.32	3.14	7.35	1.18	11.67

Treat surfaces of concrete before setting

	Unit	Hours	Hours £	Mat'ls £	O & P £	Total £
tamping	m2	0.06	0.59	0.00	0.09	0.68
floating	m2	0.10	0.98	0.00	0.15	1.13
trowelling	m2	0.15	1.47	0.00	0.22	1.69

Pavings

	Unit	Spec'list price £	O & P £	Total £
Bitumen macadam paving consisting of 50mm base course with 28mm aggregate and 20mm thick wearing course with 6mm aggregate	m2	21.05	3.16	24.21
Bitumen macadam paving consisting of 70mm base course with 40mm aggregate and 25mm thick wearing course with 10mm aggregate	m2	25.15	3.77	28.92

	Unit	Hours	Hours £	Mat'ls £	O & P £	Total £
Gravel paving in two layers, bottom layer clinker aggregate, second layer fine gravel aggregate						
50mm thick	m	0.10	0.98	2.45	0.15	3.58
75mm thick	m	0.12	1.18	3.19	0.54	4.91
100mm thick	m	0.14	1.37	4.44	0.68	6.50

	Unit	Hours	Hours £	Mat'ls £	O & P £	Total £
Brick paving, 215 x 103 x 65mm laid to falls and cross falls bedding in cement mortar 15mm thick						
straight joints both ways						
bricks laid flat	m2	1.00	10.80	23.79	1.62	36.21
bricks laid on edge	m2	1.25	13.50	36.25	5.59	55.34
herringbone pattern						
bricks laid flat	m2	1.25	13.50	23.79	2.03	39.32
bricks laid on edge	m2	1.50	16.20	36.25	6.00	58.45
Preacast concrete paving flags spot bedded in cement mortar						
natural colour						
450 × 450 × 50mm	m2	0.50	5.40	12.45	0.81	18.66
600 × 600 × 50mm	m2	0.48	5.18	10.22	2.65	18.05
600 × 900 × 50mm	m2	0.45	4.86	9.49	2.26	16.61
coloured						
450 × 450 × 50mm	m2	0.50	5.40	13.79	0.81	20.00
600 × 600 × 50mm	m2	0.48	5.18	11.07	2.85	19.10
600 × 900 × 50mm	m2	0.15	1.62	10.15	1.90	13.67

	Unit	Spec'list Price £	O & P £	Total £

ALTERATIONS AND REPAIRS

Shoring

Raking shoring consisting of two
175 × 175mm rakers, each 5m
length, 50 × 150mm wall plate
including sole pieces, steel dogs
needles, cleats and bracing

	nr	658.00	109.55	767.55

Raking shoring consisting of two
225 × 225mm rakers, each 8m
length, 50 × 200mm wall plate
including sole pieces, steel dogs
needles, cleats, bracing and
cutting through walls, floors and
ceilings

	nr	827.00	137.69	964.69

Dead shoring consisting of two
175 × 175mm legs including
needle, sole plate, braces, wedges
steel dogs and cutting through
walls, floors and ceilings

	nr	576.00	95.93	671.93

Flying shoring consisting of 100 ×
150mm shores, 50 × 200 wall
plates, 50 × 100mm struts and
straining pieces, distance between
walls being supported

3m	nr	489.00	81.41	570.41
4m	nr	523.00	87.07	610.07
5m	nr	576.00	95.90	671.90

Strut existing window openings
consisting of 125 × 50mm wall
plates and struts

	nr	123.00	20.47	143.47

	Unit	Hours	Hours £	Mat'ls £	O & P £	Total £

Forming openings

Form opening for windows or
doors in existing walls in cement
mortar and make good to sides
of openings

	Unit	Hours	Hours £	Mat'ls £	O & P £	Total £
75mm blockwork	m2	2.00	19.60	2.49	3.31	25.40
100mm blockwork	m2	2.22	21.76	2.49	3.64	27.88
140mm blockwork	m2	2.65	25.97	3.05	4.35	33.37
215mm blockwork	m2	2.80	27.44	3.66	4.67	35.77
half brick wall	m2	2.90	28.42	2.49	4.64	35.55
one brick wall	m2	3.78	37.04	3.19	6.04	46.27
one and a half brick wall	m2	4.80	47.04	3.96	7.65	58.65
two brick wall	m2	7.20	70.56	5.20	11.36	87.12

Form opening for windows or
doors in existing walls in lime
mortar and make good to sides
of openings

	Unit	Hours	Hours £	Mat'ls £	O & P £	Total £
75mm blockwork	m2	1.80	17.64	2.49	3.02	23.15
100mm blockwork	m2	2.00	19.60	2.49	3.31	25.40
140mm blockwork	m2	2.40	23.52	3.05	3.99	30.56
215mm blockwork	m2	2.55	24.99	3.66	4.30	32.95
half brick wall	m2	2.60	25.48	2.49	4.20	32.17
one brick wall	m2	3.40	33.32	3.19	5.48	41.99
one and a half brick wall	m2	4.80	47.04	3.96	7.65	58.65
two brick wall	m2	6.50	63.70	5.20	10.34	79.24

Form opening for lintels above
openings in existing walls in cement
mortar and make good to sides
of openings

	Unit	Hours	Hours £	Mat'ls £	O & P £	Total £
75mm blockwork	m2	2.98	29.20	1.11	4.55	34.86
100mm blockwork	m2	3.36	32.93	1.11	5.11	39.14
140mm blockwork	m2	3.87	37.93	1.77	5.95	45.65
215mm blockwork	m2	4.23	41.45	2.44	6.58	50.48
half brick wall	m2	6.30	61.74	1.77	9.53	73.04

	Unit	Hours	Hours £	Mat'ls £	O & P £	Total £
one brick wall	m2	10.50	102.90	3.19	15.91	122.00
one and a half brick wall	m2	12.60	123.48	3.96	19.12	146.56
two brick wall	m2	16.98	166.40	5.20	25.74	197.34

Form opening for lintels above
openings in existing walls in lime
mortar and make good to sides
of openings

	Unit	Hours	Hours £	Mat'ls £	O & P £	Total £
75mm blockwork	m2	2.75	26.95	2.49	4.42	33.86
100mm blockwork	m2	3.00	29.40	2.49	4.78	36.67
140mm blockwork	m2	3.45	33.81	3.05	5.53	42.39
215mm blockwork	m2	3.86	37.83	3.66	6.22	47.71
half brick wall	m2	5.75	56.35	2.49	8.83	67.67
one brick wall	m2	9.45	92.61	3.19	14.37	110.17
one and a half brick wall	m2	11.40	111.72	3.96	17.35	133.03
two brick wall	m2	15.30	149.94	5.20	23.27	178.41

Form opening in reinforced
concrete floor slab and make good
to edges, slab thickness

	Unit	Hours	Hours £	Mat'ls £	O & P £	Total £
100mm	m2	5.65	55.37	2.77	8.72	66.86
150mm	m2	7.86	77.03	3.05	12.01	92.09
200mm	m2	9.50	93.10	3.33	14.46	110.89
250mm	m2	11.57	113.39	3.60	17.55	134.53
300mm	m2	14.55	142.59	3.88	21.97	168.44

Form opening in reinforced
concrete walls and make good
to edges, wall thickness

	Unit	Hours	Hours £	Mat'ls £	O & P £	Total £
100mm	m2	6.50	63.70	2.77	9.97	76.44
150mm	m2	7.20	70.56	3.05	11.04	84.65
200mm	m2	10.33	101.23	3.33	15.68	120.25
250mm	m2	12.10	118.58	3.60	18.33	140.51
300mm	m2	15.32	150.14	3.88	23.10	177.12

	Unit	Hours	Hours £	Mat'ls £	O & P £	Total £

Filling openings

Take out existing door and frame
complete, make good to reveals
and piece up skirting to match
existing

	Unit	Hours	Hours £	Mat'ls £	O & P £	Total £
single internal door	nr	6.00	58.80	10.96	10.46	80.22
single external door	nr	6.20	60.76	10.96	10.76	82.48
double internal door	nr	6.20	60.76	11.58	10.85	83.19
double external door	nr	6.40	62.72	11.58	11.15	85.45

Take out existing single door and
frame complete, fill in opening, fix
new skirting to match existing and
plaster both sides

	Unit	Hours	Hours £	Mat'ls £	O & P £	Total £
100mm blockwork	m2	10.45	102.41	39.73	21.32	163.46
215mm blockwork	m2	12.47	122.21	58.07	27.04	207.32
half brick wall	m2	12.40	121.52	36.29	23.67	181.48
one brick wall	m2	14.60	143.08	65.45	31.28	239.81
one and a half brick wall	m2	16.37	160.43	90.18	37.59	288.20

Take out existing double doors and
frame complete, fill in opening, fix
new skirting to match existing and
plaster both sides

	Unit	Hours	Hours £	Mat'ls £	O & P £	Total £
100mm blockwork	m2	14.59	142.98	46.84	28.47	218.30
215mm blockwork	m2	16.36	160.33	105.09	39.81	305.23
half brick wall	m2	15.54	152.29	60.35	31.90	244.54
one brick wall	m2	20.34	199.33	110.85	46.53	356.71
one and a half brick wall	m2	22.20	217.56	168.04	57.84	443.44

	Unit	Hours	Hours £	Mat'ls £	O & P £	Total £

Underpinning

Excavate preliminary trench by
hand, maximum depth not
exceeding

	Unit	Hours	Hours £	Mat'ls £	O & P £	Total £
1.00m	m3	3.70	36.26	0.00	5.44	41.70
1.50m	m3	4.15	40.67	0.00	6.10	46.77
2.00m	m3	4.40	43.12	0.00	6.47	49.59
2.50m	m3	6.10	59.78	0.00	8.97	68.75
3.00m	m3	7.25	71.05	0.00	10.66	81.71

Excavate trench by hand below
existing foundations, maximum
not exceeding

	Unit	Hours	Hours £	Mat'ls £	O & P £	Total £
1.00m	m3	4.25	41.65	0.00	6.25	47.90
1.50m	m3	4.75	46.55	0.00	6.98	53.53
2.00m	m3	5.45	53.41	0.00	8.01	61.42
2.50m	m3	7.30	71.54	0.00	10.73	82.27
3.00m	m3	8.35	81.83	0.00	12.27	94.10

Excavate and backfill working
space, maximum depth not
exceeding

	Unit	Hours	Hours £	Mat'ls £	O & P £	Total £
1.00m	m3	5.00	49.00	0.00	7.35	56.35
1.50m	m3	5.25	51.45	0.00	7.72	59.17
2.00m	m3	5.90	57.82	0.00	8.67	66.49
2.50m	m3	8.10	79.38	0.00	11.91	91.29
3.00m	m3	9.25	90.65	0.00	13.60	104.25

Open-boarded earthwork support
to sides of preliminary trenches,
distance between faces not
exceeding 2.0m

	Unit	Hours	Hours £	Mat'ls £	O & P £	Total £
maximum depth not exceeding 1.0m	m2	0.30	2.94	1.72	0.70	5.36

	Unit	Hours	Hours £	Mat'ls £	O & P £	Total £
maximum depth not exceeding 2.0m	m2	0.40	3.92	1.72	0.85	6.49

Close-boarded earthwork support
to sides of preliminary trenches,
distance between faces not
exceeding 2.0m

	Unit	Hours	Hours £	Mat'ls £	O & P £	Total £
maximum depth not exceeding 1.0m	m2	0.85	8.33	2.93	1.69	12.95
maximum depth not exceeding 2.0m	m2	1.10	10.78	2.93	2.06	15.77

Open-boarded earthwork support
to sides of excavation trenches,
distance between faces not
exceeding 2.0m

	Unit	Hours	Hours £	Mat'ls £	O & P £	Total £
maximum depth not exceeding 1.0m	m2	0.35	3.43	1.72	0.77	5.92
maximum depth not exceeding 2.0m	m2	0.45	4.41	1.72	0.92	7.05

Close-boarded earthwork support
to sides of excavation trenches,
distance between faces not
exceeding 2.0m

	Unit	Hours	Hours £	Mat'ls £	O & P £	Total £
maximum depth not exceeding 1.0m	m2	0.95	9.31	2.93	1.84	14.08
maximum depth not exceeding 2.0m	m2	1.20	11.76	2.93	2.20	16.89

	Unit	Hours	Hours £	Mat'ls £	O & P £	Total £
Cut away projecting concrete foundations, size						
150 × 150mm	m	0.45	4.41	0.00	0.66	5.07
150 × 225mm	m	0.55	5.39	0.00	0.81	6.20
150 × 300mm	m	0.65	6.37	0.00	0.96	7.33
Cut away projecting brickwork in footings, one brick thick						
one course	m	1.10	10.78	0.00	1.62	12.40
two courses	m	1.90	18.62	0.00	2.79	21.41
three courses	m	2.60	25.48	0.00	3.82	29.30
Prepare underside of existing foundations to receive the new work, width						
300mm	m	0.75	7.35	0.00	1.10	8.45
500mm	m	1.15	11.27	0.00	1.69	12.96
750mm	m	1.40	13.72	0.00	2.06	15.78
1000mm	m	1.60	15.68	0.00	2.35	18.03
1200mm	m	1.80	17.64	0.00	2.65	20.29
Load surplus excavated material into barrows, wheel and deposit in temporary spoil heaps, average distance						
15m	m3	1.25	12.25	0.00	1.84	14.09
25m	m3	1.45	14.21	0.00	2.13	16.34
50m	m3	1.70	16.66	0.00	2.50	19.16
Load surplus excavated material into barrows, wheel and deposit in temporary spoil heaps, average distance						
25m	m3	1.45	14.21	0.00	2.13	16.34
50m	m3	1.70	16.66	0.00	2.50	19.16

	Unit	Hours	Hours £	Mat'ls £	O & P £	Total £
Load surplus excavated material into barrows, wheel and deposit in skips or lorries, average distance						
25m	m3	1.40	13.72	0.00	2.06	15.78
50m	m3	1.65	16.17	0.00	2.43	18.60
Level and compact bottom of excavation	m2	0.15	1.47	0.00	0.22	1.69
Site-mixed concrete 1:3:6 40mm aggregate in foundations to under-pinning, thickness						
150 to 300mm	m3	4.35	42.63	68.98	16.74	128.35
300 to 450mm	m3	4.05	39.69	68.98	16.30	124.97
over 450mm	m3	3.65	35.77	68.98	15.71	120.46
Site-mixed concrete 1:2:4 20mm aggregate in foundations to under-pinning, thickness						
150 to 300mm	m3	4.35	42.63	77.11	17.96	137.70
300 to 450mm	m3	4.05	39.69	77.11	17.52	134.32
over 450mm	m3	3.65	35.77	77.11	16.93	129.81
Plain vertical formwork to sides of underpinned foundations, height						
over 1m	m2	2.30	22.54	8.85	4.71	36.10
not exceeding 250mm	m	0.75	7.35	8.85	2.43	18.63
250 to 500mm	m	1.25	12.25	8.85	3.17	24.27
500mm to 1m	m	1.70	16.66	8.85	3.83	29.34

	Unit	Hours	Hours £	Mat'ls £	O & P £	Total £
Plain vertical formwork to sides of underpinned foundations, left in, height						
over 1m	m2	2.15	21.07	21.53	6.39	48.99
not exceeding 250mm	m	0.65	6.37	7.65	2.10	16.12
250 to 500mm	m	1.15	11.27	14.16	3.81	29.24
500mm to 1m	m	1.60	15.68	21.53	5.58	42.79
High yield deformed steel reinforcement bars, straight or bent						
10mm diameter	m	0.04	0.39	0.28	0.10	0.77
12mm diameter	m	0.05	0.49	0.37	0.13	0.99
16mm diameter	m	0.06	0.59	0.59	0.18	1.35
20mm diameter	m	0.07	0.69	0.95	0.25	1.88
25mm diameter	m	0.08	0.78	1.55	0.35	2.68
Common bricks basic price £200 per thousand in cement mortar in underpinning						
one brick thick	m2	5.20	62.40	36.85	14.89	114.14
one and a half brick thick	m2	7.40	88.80	55.27	21.61	165.68
two brick thick	m2	9.35	112.20	95.90	31.22	239.32
Class A engineering bricks in basic price £350 per thousand in cement mortar in underpinning						
one brick thick	m2	5.40	64.80	63.55	19.25	147.60
one and a half brick thick	m2	7.60	91.20	95.33	27.98	214.51
two brick thick	m2	9.50	114.00	127.17	36.18	277.35
Hessian-based bitumen damp proof course, bedded in cement mortar, horizontal						
over 225mm wide	m2	0.35	4.20	9.03	1.98	15.21
112mm wide	m	0.05	0.60	1.05	0.25	1.90

	Unit	Hours	Hours £	Mat'ls £	O & P £	Total £
Two courses of slates bedded in cement mortar, horizontal						
over 225mm wide	m2	2.90	34.80	30.94	9.86	75.60
225mm wide	m	0.85	10.20	6.30	2.48	18.98
Wedge and pin new work to soffit of existing with slates in cement mortar, width						
one brick thick	m	1.90	22.80	6.30	4.37	33.47
one and a half brick thick	m	2.40	28.80	8.78	5.64	43.22
two brick thick	m	2.80	33.60	12.95	6.98	53.53

Temporary screens

Erect, maintain and remove temporary screens consisting of 50 × 50mm softwood framing covered one side with

	Unit	Hours	Hours £	Mat'ls £	O & P £	Total £
hardboard (three uses)	m2	1.15	13.80	1.33	2.27	17.40
insulation board (three uses)	m2	1.15	13.80	1.88	2.35	18.03
polythene sheeting (three uses)	m2	0.80	9.60	0.13	1.46	11.19

SPOT ITEMS

Brickwork, blockwork and masonry

Take out existing fireplace, fill opening with 100mm thick block-work plastered one side, fix new skirting to match existing, make good flooring where hearth removed

	Unit	Hours	Hours £	Mat'ls £	O & P £	Total £
medium size fireplace	nr	12.00	144.00	19.48	24.52	188.00
large size fireplace	nr	14.00	168.00	24.07	28.81	220.88

	Unit	Hours	Hours £	Mat'ls £	O & P £	Total £
Cut out existing projecting chimney breast, size 1350 × 350mm, including making good flooring and ceiling to match existing and plastering wall where breast removed						
one storey	nr	36.00	432.00	38.29	70.54	540.83
Take down chimney stack to 100mm below roof level, seal flue with slates in cement mortar, make good roof timbers						
slated roof	nr	46.00	552.00	97.18	97.38	746.56
tiled roof	nr	46.00	552.00	62.82	92.22	707.04
Cut out single brick in half brick wall and replace in gauged mortar						
commons	nr	0.25	3.00	0.22	0.48	3.70
facings	nr	0.30	3.60	0.44	0.61	4.65
Cut out decayed brickwork in half brick wall in areas 0.5 to 1m2 and replace in gauged mortar						
commons	nr	5.20	62.40	12.32	11.21	85.93
facings	nr	6.00	72.00	34.63	15.99	122.62
Cut out decayed brickwork in one brick wall in areas 0.5 to 1m2 and replace in gauged mortar						
commons	nr	7.60	91.20	24.64	17.38	133.22
facings	nr	8.80	105.60	43.06	22.30	170.96

	Unit	Hours	Hours £	Mat'ls £	O & P £	Total £
Cut out vertical, horizontal or stepped cracks in half brick wall, replace average 350mm wide with new bricks in gauged mortar						
commons	m	2.75	33.00	4.60	5.64	43.24
facings	m	3.10	37.20	11.69	7.33	56.22
Cut out vertical, horizontal or stepped cracks in one brick wall, replace average 350mm wide with new bricks in gauged mortar						
commons	m	5.20	62.40	9.21	10.74	82.35
facings	m	5.80	69.60	23.35	13.94	106.89
Cut out defective brick-on-end soldier arch to half brick wall, and replace with new bricks in gauged mortar						
commons	m	3.00	36.00	6.99	6.45	49.44
facings	m	3.40	40.80	10.68	7.72	59.20
Cut out defective brick-on-end soldier arch to one brick wall, and replace with new bricks in gauged mortar						
commons	m	3.80	45.60	13.98	8.94	68.52
facings	m	4.20	50.40	21.37	10.77	82.54
Cut out defective terracotta air brick and replace						
215 × 65mm	nr	0.35	4.20	2.16	0.95	7.31
215 × 140mm	nr	0.50	6.00	2.90	1.34	10.24
215 × 215mm	nr	0.65	7.80	8.04	2.38	18.22

	Unit	Hours	Hours £	Mat'ls £	O & P £	Total £
Rake out joints in gauged mortar, refix loose flashing and point up on completion						
horizontal	m	0.45	5.40	0.44	0.88	6.72
stepped	m	0.65	7.80	0.66	1.27	9.73
Rake out joints and point up in gauged mortar	m2	0.65	7.80	0.66	1.27	9.73
Cut out one course of half brick wall, insert hessian-based damp course 112mm wide and replace with new bricks in gauged mortar						
commons	m	2.00	24.00	2.81	4.02	30.83
facings	m	2.00	24.00	3.19	4.08	31.27
Cut out one course of one brick wall, insert hessian-based damp course 112mm wide and replace with new bricks in gauged mortar						
commons	m	3.25	39.00	5.63	6.69	51.32
facings	m	3.25	39.00	6.39	6.81	52.20
Cut out single block, replace with new in gauged mortar, thickness						
100mm	nr	0.30	3.60	1.22	0.72	5.54
140mm	nr	0.40	4.80	1.94	1.01	7.75
190mm	nr	0.50	6.00	2.44	1.27	9.71
255mm	nr	0.60	7.20	3.30	1.58	12.08
Rake out joints of random rubble walling and point up in gauged mortar						
flush pointing	m2	0.50	6.00	0.55	0.98	7.53
weather pointing	m2	0.55	6.60	0.55	1.07	8.22

	Unit	Hours	Hours £	Mat'ls £	O & P £	Total £

Roofing

Take up roof coverings from
pitched roof

tiles	m2	0.80	7.84	0.00	1.18	9.02
slates	m2	0.80	7.84	0.00	1.18	9.02
timber boarding	m2	1.00	9.80	0.00	1.47	11.27
metal sheeting	m2	0.20	1.96	0.00	0.29	2.25
flat sheeting	m2	0.30	2.94	0.00	0.44	3.38
corrugated sheeting	m2	0.30	2.94	0.00	0.44	3.38
underfelt	m2	0.10	0.98	0.00	0.15	1.13

Take up roof coverings from
flat roof

bituminous felt	m2	0.25	2.45	0.00	0.37	2.82
metal sheeting	m2	0.30	2.94	0.00	0.44	3.38
woodwool slabs	m2	0.50	4.90	0.00	0.74	5.64
firrings	m2	0.20	1.96	0.00	0.29	2.25

Take up roof coverings from
pitched roof, carefully lay aside
for reuse

tiles	m2	1.10	10.78	0.00	1.62	12.40
slates	m2	1.10	10.78	0.00	1.62	12.40
metal sheeting	m2	0.50	4.90	0.00	0.74	5.64
flat sheeting	m2	0.60	5.88	0.00	0.88	6.76
corrugated sheeting	m2	0.60	5.88	0.00	0.88	6.76

Take up roof coverings from
flat roof, carefully lay aside for
reuse

metal sheeting	m2	0.60	5.88	0.00	0.88	6.76
woodwool slabs	m2	0.80	7.84	0.00	1.18	9.02

	Unit	Hours	Hours £	Mat'ls £	O & P £	Total £
Inspect roof battens, refix loose and replace with new, size 38 × 25mm						
25% of area						
250mm centres	m2	0.14	1.68	0.33	0.30	2.31
450mm centres	m2	0.12	1.44	0.27	0.26	1.97
600mm centres	m2	0.10	1.20	0.22	0.21	1.63
50% of area						
250mm centres	m2	0.26	3.12	0.66	0.57	4.35
450mm centres	m2	0.16	1.92	0.55	0.37	2.84
600mm centres	m2	0.18	2.16	0.44	0.39	2.99
75% of area						
250mm centres	m2	0.36	4.32	0.99	0.80	6.11
450mm centres	m2	0.26	3.12	0.83	0.59	4.54
600mm centres	m2	0.22	2.64	0.66	0.50	3.80
100% of area						
250mm centres	m2	0.44	5.28	1.33	0.99	7.60
450mm centres	m2	0.32	3.84	1.11	0.74	5.69
600mm centres	m2	0.38	4.56	0.88	0.82	6.26
Remove single slipped slate and refix	nr	1.00	12.00	0.88	1.93	14.81
Remove single broken slate, renew with new Welsh blue slate						
405 × 255mm	nr	1.20	14.40	1.33	2.36	18.09
510 × 255mm	nr	1.20	14.40	2.55	2.54	19.49
610 × 305mm	nr	1.20	14.40	5.38	2.97	22.75
Remove slates in area approximately 1m2 and replace with Welsh blue slates previously laid aside	nr	1.60	19.20	0.00	2.88	22.08

	Unit	Hours	Hours £	Mat'ls £	O & P £	Total £
Remove slates in area approximately 1m2 and replace with Welsh blue slates previously laid aside						
405 × 255mm	nr	1.80	21.60	52.41	11.10	85.11
510 × 255mm	nr	1.70	20.40	54.62	11.25	86.27
610 × 305mm	nr	1.60	19.20	58.15	11.60	88.95
Remove double course at eaves and fix new Welsh blue slates						
405 × 255mm	m	0.70	8.40	24.57	4.95	37.91
510 × 255mm	m	0.70	8.40	30.72	5.87	44.99
610 × 305mm	m	0.70	8.40	34.67	6.46	49.53
Remove single verge undercloak course and renew						
405 × 255mm	m	0.90	10.80	18.39	4.38	33.57
510 × 255mm	m	0.90	10.80	19.51	4.55	34.86
610 × 305mm	m	0.90	10.80	21.38	4.83	37.01
Remove single slipped tile and refix	nr	0.30	3.60	0.00	0.54	4.14
Remove single broken tile and renew						
Marley plain tile	nr	0.30	3.60	0.53	0.62	4.75
Marley Ludlow Plus tile	nr	0.30	3.60	0.79	0.66	5.05
Marley Modern tile	nr	0.30	3.60	1.22	0.72	5.54
Redland Renown tile	nr	0.30	3.60	1.22	0.72	5.54
Redland Norfolk tile	nr	0.30	3.60	1.05	0.70	5.35
Remove tiles in area approximately 1m2 and replace with tiles previously laid aside						
Marley plain tile	nr	1.80	21.60	27.26	7.33	56.19
Marley Ludlow Plus tile	nr	1.20	14.40	11.94	3.95	30.29
Marley Modern tile	nr	1.10	13.20	12.76	3.89	29.85

	Unit	Hours	Hours £	Mat'ls £	O & P £	Total £
Redland Renown tile	nr	1.10	13.20	12.63	3.87	29.70
Redland Norfolk tile	nr	1.15	13.80	15.56	4.40	33.76
Take off defective ridge capping and refix including pointing in mortar	m	1.10	13.20	1.46	2.20	16.86

Carpentry and joinery

Take down, cut out or demolish structural timbers and load into skips

	Unit	Hours	Hours £	Mat'ls £	O & P £	Total £
structural timbers						
50 × 100mm	m	0.10	1.20	0.00	0.18	1.38
50 × 150mm	m	0.12	1.44	0.00	0.22	1.66
75 × 100mm	m	0.14	1.68	0.00	0.25	1.93
75 × 150mm	m	0.16	1.92	0.00	0.29	2.21
100 × 150mm	m	0.18	2.16	0.00	0.32	2.48
100 × 200mm	m	0.20	2.40	0.00	0.36	2.76
roof boarding	m2	0.28	3.36	0.00	0.50	3.86
floor boarding	m2	0.22	2.64	0.00	0.40	3.04
stud partition plasterboard both sides	m2	0.75	9.00	0.00	1.35	10.35
skirtings and grounds						
100mm high	m	0.08	0.96	0.00	0.14	1.10
150mm high	m	0.09	1.08	0.00	0.16	1.24
200mm high	m	0.10	1.20	0.00	0.18	1.38
rails						
50mm high	m	0.05	0.60	0.00	0.09	0.69
75mm high	m	0.06	0.72	0.00	0.11	0.83
100mm high	m	0.07	0.84	0.00	0.13	0.97
fittings						
wall cupboards	nr	0.25	3.00	0.00	0.45	3.45
floor units	nr	0.20	2.40	0.00	0.36	2.76
sink units	nr	0.25	3.00	0.00	0.45	3.45
staircase, 900mm wide						
straight flight	nr	4.00	48.00	0.00	7.20	55.20
landing	nr	1.50	18.00	0.00	2.70	20.70

	Unit	Hours	Hours £	Mat'ls £	O & P £	Total £
doors, frames and linings						
single, internal	nr	0.40	4.80	0.00	0.72	5.52
single, external	nr	0.60	7.20	0.00	1.08	8.28
double, internal	nr	0.60	7.20	0.00	1.08	8.28
double, external	nr	0.80	9.60	0.00	1.44	11.04
windows						
casement, 1200 × 900mm	nr	0.50	6.00	0.00	0.90	6.90
casement, 1800 × 900mm	nr	0.60	7.20	0.00	1.08	8.28
sash, 900 × 1500mm	nr	0.80	9.60	0.00	1.44	11.04
sash, 1800 × 900mm	nr	0.90	10.80	0.00	1.62	12.42
ironmongery						
bolt	nr	0.20	2.40	0.00	0.36	2.76
deadlock	nr	0.25	3.00	0.00	0.45	3.45
mortice lock	nr	0.35	4.20	0.00	0.63	4.83
mortice latch	nr	0.35	4.20	0.00	0.63	4.83
cylinder lock	nr	0.25	3.00	0.00	0.45	3.45
door closer	nr	0.35	4.20	0.00	0.63	4.83
casement stay	nr	0.15	1.80	0.00	0.27	2.07
casement fastener	nr	0.15	1.80	0.00	0.27	2.07
toilet roll holder	nr	0.15	1.80	0.00	0.27	2.07
shelf bracket	nr	0.15	1.80	0.00	0.27	2.07
Cut out defective joists or rafters and replace with new						
50 × 75mm	nr	0.35	4.20	1.28	0.82	6.30
50 × 100mm	nr	0.40	4.80	1.33	0.92	7.05
50 × 150mm	nr	0.60	7.20	1.99	1.38	10.57
75 × 100mm	nr	0.65	7.80	2.68	1.57	12.05
75 × 150mm	nr	0.75	9.00	3.26	1.84	14.10
Cut out defective skirting and renew						
75mm high	m	0.30	3.60	2.10	0.86	6.56
100mm high	m	0.35	4.20	2.46	1.00	7.66
150mm high	m	0.40	4.80	3.18	1.20	9.18

	Unit	Hours	Hours £	Mat'ls £	O & P £	Total £
Ease, adjust and oil						
door	nr	0.60	7.20	0.00	1.08	8.28
casement window	nr	0.40	4.80	0.00	0.72	5.52
sash window including renewing cords	nr	1.50	18.00	4.68	3.40	26.08
Take up defective flooring and replace with 25mm thick plain edged boarding						
areas less than 1m2	m2	1.50	18.00	12.76	4.61	35.37
areas more than 1m2	m2	1.20	14.40	12.76+	2.16	16.56
Refix loose floorboards including punching in protruding nails	m2	0.20	2.40	0.00	0.36	2.76
Take down existing door, lay aside, piece up frame or lining where butts removed, rehang door to opposite hand on existing butts and hardware						
single, internal	nr	1.50	18.00	1.66	2.95	22.61
single, e×ternal	nr	1.70	20.40	1.66	3.31	25.37
Finishings						
Take down plasterboard sheeting from						
studded walls	m2	0.35	4.20	0.00	0.63	4.83
ceilings	m2	0.40	4.80	0.00	0.72	5.52
Hack off plaster from						
walls	m2	0.25	3.00	0.00	0.45	3.45
ceiling	m2	0.30	3.60	0.00	0.54	4.14

	Unit	Hours	Hours £	Mat'ls £	O & P £	Total £
Make good plaster to walls where wall removed						
100mm wide	m2	0.95	11.40	1.66	1.96	15.02
150mm wide	m2	1.05	12.60	1.77	2.16	16.53
200mm wide	m2	1.15	13.80	1.88	2.35	18.03
Make good plaster to ceilings where wall removed						
100mm wide	m2	1.10	13.20	1.66	2.23	17.09
150mm wide	m2	1.15	13.80	1.77	2.34	17.91
200mm wide	m2	1.20	14.40	1.88	2.44	18.72
Cut out cracks in plasterwork and make good						
walls	m2	0.35	4.20	0.27	0.67	5.14
ceiling	m2	0.40	4.80	0.27	0.76	5.83
Hack off wall tiling and make good surface	m2	0.75	9.00	5.06	2.11	16.17

Plumbing

	Unit	Hours	Hours £	Mat'ls £	O & P £	Total £
Remove sanitary fittings and supports, seal off supply and waste pipes and prepare to receive new fittings						
bath	nr	3.50	50.75	11.10	9.28	71.13
sink	nr	3.50	50.75	11.10	9.28	71.13
lavatory basin	nr	3.25	47.13	6.66	8.07	61.85
WC	nr	3.75	54.38	6.66	9.16	70.19
shower cubicle	nr	4.00	58.00	8.88	10.03	76.91
bidet	nr	3.25	47.13	6.66	8.07	61.85

	Unit	Hours	Hours £	Mat'ls £	O & P £	Total £
Take down length of existing gutter, prepare ends and install new length of gutter to existing brackets						
PVC-U						
76mm	nr	0.60	7.20	7.88	2.26	17.34
112mm	nr	0.65	7.80	9.65	2.62	20.07
cast iron						
100mm	nr	1.12	13.44	23.88	5.60	42.92
Take down existing brackets from fascias and replace with galvanised steel repair brackets at 1m centres	m	0.30	3.60	2.38	0.90	6.88
Take down length of existing pipe, prepare ends and install new length of pipe						
PVC-U						
68mm diameter	nr	0.75	9.00	8.73	2.66	20.39
68mm square	nr	0.75	9.00	9.59	2.79	21.38
cast iron						
75mm	nr	0.90	10.80	35.31	6.92	53.03
100mm	nr	1.05	12.60	48.70	9.20	70.50
Take down fittings and install new						
PVC-U						
68mm diameter bend	nr	0.50	7.25	3.26	1.58	12.09
68mm offset	nr	0.50	7.25	7.60	2.23	17.08
68mm branch	nr	0.50	7.25	7.77	2.25	17.27
68mm shoe	nr	0.75	10.88	7.02	2.68	20.58
Cast iron						
75mm diameter bend	nr	0.60	8.70	6.54	2.29	17.53
75mm offset	nr	0.60	8.70	16.51	3.78	28.99
75mm branch	nr	0.60	8.70	7.18	2.38	18.26
75mm shoe	nr	0.60	8.70	15.49	3.63	27.82

	Unit	Hours	Hours £	Mat'ls £	O & P £	Total £
Cast iron						
100mm diameter bend	nr	0.70	10.15	22.00	4.82	36.97
100mm offset	nr	0.70	10.15	30.45	6.09	46.69
100mm branch	nr	0.70	10.15	8.79	2.84	21.78
100mm shoe	nr	0.70	10.15	18.53	4.30	32.98
Cut out 500mm length of copper pipe, install new pipe with compression fittings at each end						
15mm	nr	0.80	11.60	4.30	2.39	18.29
22mm	nr	0.90	13.05	7.54	3.09	23.68
Take off existing radiator valve and replace with new	nr	0.90	13.05	9.83	3.43	26.31
Take out existing galvanised steel water storage tank and install new plastic tank complete with ball valve, lid and insulation including cutting holes, make up pipework and connectors to existing pipework						
PC4, 16 litres	nr	3.60	52.20	15.80	10.20	78.20
PC15, 68 litres	nr	3.70	53.65	40.67	14.15	108.47
PC20, 91 litres	nr	3.75	54.38	42.56	14.54	111.48
PC25, 114 litres	nr	3.80	55.10	53.23	16.25	124.58
PC40, 182 litres	nr	4.90	71.05	85.24	23.44	179.73
PC50, 227 litres	nr	5.00	72.50	90.89	24.51	187.90

Glazing

	Unit	Hours	Hours £	Mat'ls £	O & P £	Total £
Hack out glass and remove	m2	0.45	5.40	0.00	0.81	6.21
Clean rebates, remove sprigs or clips and prepare for reglazing	m	0.20	2.40	0.00	0.36	2.76

	Unit	Hours	Hours £	Mat'ls £	O & P £	Total £

Painting

Prepare, wash down painted
surfaces, rub down to receive new
paintwork

brickwork	m2	0.14	1.68	0.00	0.25	1.93
blockwork	m2	0.15	1.80	0.00	0.27	2.07
plasterwork	m2	0.12	1.21	0.00	0.18	1.40

Prepare, wash down previously
painted wood surfaces, rub down
to receive new paintwork surfaces

over 300mm girth	m2	0.28	3.36	0.00	0.50	3.86
isolated surfaces not exceeding 300mm girth	m	0.10	1.20	0.00	0.18	1.38
isolated surfaces not exceeding 0.5m2	nr	0.12	1.44	0.00	0.22	1.66

Wallpapering

Strip off one layer of existing
paper, stop cracks and rub down
to receive new paper

woodchip

walls	m2	0.18	2.16	0.00	0.32	2.48
walls in staircase areas	m2	0.20	2.40	0.00	0.36	2.76
ceilings	m2	0.22	2.64	0.00	0.40	3.04
ceilings in staircase areas	m2	0.24	2.88	0.00	0.43	3.31

vinyl

walls	m2	0.23	2.76	0.00	0.41	3.17
walls in staircase areas	m2	0.25	3.00	0.00	0.45	3.45
ceilings	m2	0.27	3.24	0.00	0.49	3.73
ceilings in staircase areas	m2	0.29	2.93	0.00	0.44	3.37

	Unit	Hours	Hours £	Mat'ls £	O & P £	Total £
standard patterned						
walls	m2	0.21	2.52	0.00	0.38	2.90
walls in staircase areas	m2	0.23	2.76	0.00	0.41	3.17
ceilings	m2	0.25	3.00	0.00	0.45	3.45
ceilings in staircase areas	m2	0.27	3.24	0.00	0.49	3.73

Part Two

DAMAGE REPAIRS

Emergency measures

Fire damage

Flood damage

Gale damage

Theft damage

	Unit	Hours	Hours £	Plant £	O & P £	Total £

EMERGENCY MEASURES

Flooding

	Unit	Hours	Hours £	Plant £	O & P £	Total £
Instal hired pump and hoses in position	Item	1.00	12.00	0.00	1.80	13.80
Hire diaphragm pump and hoses						
50mm	hour	1.00	0.00	3.20	0.48	3.68
75mm	hour	1.00	0.00	3.47	0.52	3.99

	Unit	Hours	Hours £	Mat'ls £	O & P £	Total £
Hoardings and screens						
Temporary screens and hoardings 2m high consisting of 22mm thick exterior quality plywood fixed to 100 × 50mm posts and rails	m	2.60	31.20	31.15	9.35	71.70
Hire tarpaulin sheeting size 5 × 4m fixed in position	day	0.00	0.00	3.81	0.57	4.38
Plywood sheeting blocking up window or door opening	m2	1.00	12.00	15.58	4.14	31.72
Shoring						
Timber dead shores consisting of 200 × 200mm shores, 250 × 50mm plates and 200 × 50mm braces at centres of						
2m	m2	3.80	45.60	73.24	17.83	136.67
3m	m2	3.00	36.00	57.88	14.08	107.96
4m	m2	2.20	26.40	41.68	10.21	78.29

	Unit	Hours	Hours £	Plant £	O & P £	Total £
Timber raking and flying shores consisting of 200 × 200mm shores, 250 × 50mm plates and 200 × 50mm braces to gable end of two-storey house	Item	72.00	864.00	520.40	207.66	1592.06

Access towers

Hire narrow width access tower
size 0.85 × 1.8m, height

	Unit	Hours	Hours £	Plant £	O & P £	Total £
7.20m	week	1.00	0.00	111.28	16.69	127.97
6.20m	week	1.00	0.00	92.02	13.80	105.82
5.20m	week	1.00	0.00	77.04	11.56	88.60
4.20m	week	1.00	0.00	62.06	9.31	71.37

Hire double width access tower
size 1.45 × 2.5m, height

	Unit	Hours	Hours £	Plant £	O & P £	Total £
7.20m	week	1.00	0.00	128.40	19.26	147.66
6.20m	week	1.00	0.00	119.84	17.98	137.82
5.20m	week	1.00	0.00	107.00	16.05	123.05
4.20m	week	1.00	0.00	89.88	13.48	103.36

	Unit	Hours	Hours £	Mat'ls £	O & P £	Total £

FIRE DAMAGE

Window replacement

Take out existing window, prepare
jambs and cill to receive new
PVC-U

	Unit	Hours	Hours £	Mat'ls £	O & P £	Total £
size 600 × 900mm	nr	2.00	24.00	118.32	21.35	163.67
size 600 × 1200mm	nr	2.00	24.00	156.63	27.09	207.72
size 1200 × 1200mm	nr	2.50	30.00	180.76	31.61	242.37
size 1800 × 1200mm	nr	3.00	36.00	191.05	34.06	261.11

softwood painted

size 630 × 900mm	nr	2.00	24.00	84.13	16.22	124.35
size 915 × 900mm	nr	2.00	24.00	90.24	17.14	131.38
size 915 × 1200mm	nr	2.50	30.00	95.62	18.84	144.46
size 1200 × 1200mm	nr	3.00	36.00	112.85	22.33	171.18

hardwood stained

size 915 × 1050mm	nr	2.00	24.00	185.40	31.41	240.81
size 915 × 1500mm	nr	2.00	19.00	205.80	33.72	258.52
size 1200 × 1500mm	nr	2.50	23.75	220.24	36.60	280.59
size 1770 × 1200mm	nr	3.00	28.50	263.63	43.82	335.95

Take out existing bay window,
prepare jambs and cill to receive
new
PVC-U

size 1800 × 900mm	nr	3.00	36.00	215.58	37.74	289.32
size 1800 × 1200mm	nr	3.00	36.00	241.56	41.63	319.19
size 2400 × 900mm	nr	3.50	42.00	414.36	68.45	524.81
size 2400 × 1200mm	nr	3.50	42.00	442.62	72.69	557.31

softwood painted

size 1800 × 900mm	nr	3.00	36.00	153.61	28.44	218.05
size 1800 × 1200mm	nr	3.00	36.00	177.48	32.02	245.50
size 2400 × 900mm	nr	3.50	42.00	264.88	46.03	352.91
size 2400 × 1200mm	nr	3.50	42.00	288.54	49.58	380.12

	Unit	Hours	Hours £	Mat'ls £	O & P £	Total £
hardwood stained						
size 1800 × 900mm	nr	3.00	36.00	256.33	43.85	336.18
size 1800 × 1200mm	nr	3.00	36.00	287.09	48.46	371.55
size 2400 × 900mm	nr	3.50	42.00	327.25	55.39	424.64
size 2400 × 1200mm	nr	3.50	42.00	359.13	60.17	461.30

Window repairs

	Unit	Hours	Hours £	Mat'ls £	O & P £	Total £
Take off and replace defective ironmongery to softwood windows						
casement fastener	nr	0.20	2.40	3.90	0.95	7.25
casement stay	nr	0.20	2.40	5.93	1.25	9.58
hinges	pair	0.25	3.00	3.54	0.98	7.52
cockspur fastener	nr	0.20	2.40	9.85	1.84	14.09
sash fastener	nr	0.25	3.00	4.54	1.13	8.67
Take off and replace defective ironmongery to hardwood windows						
casement fastener	nr	0.30	3.60	3.90	1.13	8.63
casement stay	nr	0.30	3.60	5.93	1.43	10.96
hinges	pair	0.35	4.20	3.54	1.16	8.90
cockspur fastener	nr	0.30	3.60	9.87	2.02	15.49
sash fastener	nr	0.35	4.20	4.54	1.31	10.05
Take off and replace defective ironmongery to PVC-U windows						
casement fastener	nr	0.20	2.40	3.90	0.95	7.25
casement stay	nr	0.20	2.40	5.93	1.25	9.58
hinges	pair	0.25	3.00	3.54	0.98	7.52
cockspur fastener	nr	0.20	2.40	9.87	1.84	14.11
sash fastener	nr	0.25	3.00	4.54	1.13	8.67
Take out existing window cill and replace						
softwood	m	0.55	6.60	8.07	2.20	16.87
hardwood	m	0.65	7.80	23.92	4.76	36.48

	Unit	Hours	Hours £	Mat'ls £	O & P £	Total £
Take out existing window board and replace						
softwood	m	0.45	5.40	6.20	1.74	13.34
hardwood	m	0.60	7.20	19.97	4.08	31.25

Door replacement

	Unit	Hours	Hours £	Mat'ls £	O & P £	Total £
Take off existing external door and frame and replace with new flush door						
softwood	nr	2.20	26.40	93.49	17.98	137.87
hardwood	nr	2.50	30.00	336.26	54.94	421.20
PVC-U	nr	2.20	26.40	309.56	50.39	386.35
panelled door						
softwood	nr	2.20	26.40	117.82	21.63	165.85
hardwood	nr	2.50	30.00	368.67	59.80	458.47
PVC-U	nr	2.20	26.40	345.97	55.86	428.23
half glazed door						
softwood	nr	2.20	26.40	117.82	21.63	165.85
hardwood	nr	2.50	30.00	368.67	59.80	458.47
PVC-U	nr	2.20	26.40	345.97	55.86	428.23
fully glazed door						
softwood	nr	2.80	33.60	94.52	19.22	147.34
hardwood	nr	3.20	38.40	348.43	58.02	444.85
PVC-U	nr	2.80	33.60	321.54	53.27	408.41
fully glazed patio doors						
softwood	pair	3.80	45.60	212.11	38.66	296.37
hardwood	pair	4.30	51.60	489.40	81.15	622.15
PVC-U	pair	3.80	45.60	451.68	74.59	571.87
galvanised steel up-and-over garage doors						
2135 × 1980mm	nr	6.00	72.00	272.03	51.60	395.63
3965 × 2135mm	nr	7.50	90.00	951.39	156.21	1197.60

	Unit	Hours	Hours £	Mat'ls £	O & P £	Total £
Door repairs						
Take off and replace defective ironmongery to softwood doors						
bolts						
barrel	nr	0.20	2.40	5.92	1.25	9.57
flush	nr	0.20	2.40	7.72	1.52	11.64
tower	nr	0.30	3.60	7.09	1.60	12.29
butts						
light	pair	0.20	2.40	3.55	0.89	6.84
medium	pair	0.25	3.00	3.61	0.99	7.60
heavy	pair	0.30	2.85	4.54	1.11	8.50
locks						
cupboard	nr	0.30	3.60	6.99	1.59	12.18
mortice dead lock	nr	0.85	10.20	12.02	3.33	25.55
rim lock	nr	0.45	5.40	5.46	1.63	12.49
cylinder	nr	1.10	13.20	23.89	5.56	42.65
Take off and replace defective ironmongery to hardwood doors						
bolts						
barrel	nr	0.30	3.60	5.92	1.43	10.95
flush	nr	0.30	3.60	7.72	1.70	13.02
indicating	nr	0.30	3.60	7.78	1.71	13.09
tower	nr	0.40	4.80	7.09	1.78	13.67
butts						
light	pair	0.40	4.80	3.55	1.25	9.60
medium	pair	0.35	4.20	3.61	1.17	8.98
heavy	pair	0.40	4.80	4.54	1.40	10.74
locks						
cupboard	nr	0.30	3.60	6.99	1.59	12.18
mortice dead lock	nr	0.85	10.20	12.02	3.33	25.55
rim lock	nr	0.45	5.40	5.46	1.63	12.49
cylinder	nr	1.10	13.20	23.89	5.56	42.65

	Unit	Hours	Hours £	Mat'ls £	O & P £	Total £

Partitions, walls and ceilings

Pull down existing damaged
partitions and walls and rebuild

stud partition plasterboard both						
sides	m2	1.50	18.00	8.75	4.01	30.76
brickwork 112mm thick						
plastered both sides	m2	4.00	48.00	15.18	9.48	72.66
blockwork 75mm thick						
plastered both sides	m2	3.10	37.20	9.63	7.02	53.85
blockwork 100mm thick						
plastered both sides	m2	3.30	39.60	19.79	8.91	68.30

Hack off scorched plaster to walls

and renew	m2	1.50	18.00	1.81	2.97	22.78

Pull down plasterboard and skim

ceilings and renew	m2	1.56	18.72	3.28	3.30	25.30

Hack off damaged wall tiles and

renew	m2	1.45	17.40	22.68	6.01	46.09

Take off damaged skirting and
replace

19 × 75mm softwood	m	0.27	3.24	2.00	0.79	6.03
19 × 100mm softwood	m	0.30	3.60	2.32	0.89	6.81
19 × 100mm hardwood	m	0.32	3.84	5.26	1.37	10.47
25 × 150mm hardwood	m	0.38	4.56	9.88	2.17	16.61

Take up damaged flooring and
renew

19mm tongued and grooved						
softwood flooring	m2	1.14	13.68	10.05	3.56	27.29
25mm tongued and grooved						
softwood flooring	m2	1.15	13.80	11.82	3.84	29.46
12mm tongued and grooved						
chipboard flooring	m2	0.80	9.60	5.22	2.22	17.04
18mm tongued and grooved						
chipboard flooring	m2	1.00	12.00	5.62	2.64	20.26
2mm vinyl sheeting	m2	0.50	6.00	9.56	2.33	17.89

	Unit	Hours	Hours £	Mat'ls £	O & P £	Total £
2.5mm vinyl sheeting	m2	0.55	6.60	12.70	2.90	22.20
3mm vinyl sheeting	m2	0.60	7.20	12.94	3.02	23.16
2mm vinyl tiling	m2	0.40	4.80	8.33	1.97	15.10
3mm laminated wood floorir	m2	0.90	10.80	10.13	3.14	24.07
12.5mm quarry tiling	m2	1.15	13.80	30.55	6.65	51.00
Cut out and replace floor joists and roof members						
38 × 100mm sawn softwood	m	0.22	2.64	1.30	0.59	4.53
50 × 75mm sawn softwood	m	0.24	2.88	1.71	0.69	5.28
50 × 100mm sawn softwood	m	0.28	3.36	2.05	0.81	6.22
50 × 125mm sawn softwood	m	0.30	3.60	2.24	0.88	6.72
50 × 150mm sawn softwood	m	0.32	3.84	2.43	0.94	7.21
75 × 125mm sawn softwood	m	0.34	4.08	2.86	1.04	7.98
75 × 200mm sawn softwood	m	0.36	4.32	3.49	1.17	8.98
Take down damaged softwood staircase 2600mm rise and renew						
straight flight 900mm wide	nr	12.20	146.40	585.57	109.80	841.77
straight flight 900mm wide with balustrade	nr	13.30	159.60	672.89	124.87	957.36
two flights 900mm wide with landing and balustrade	nr	14.50	174.00	751.28	138.79	1064.07
Take down damaged hardwood staircase 2600mm rise and renew						
straight flight 900mm wide	nr	12.40	148.80	993.29	171.31	1313.40
straight flight 900mm wide with balustrade	nr	13.90	166.80	1083.47	187.54	1437.81
two flights 900mm wide with landing and balustrade	nr	15.00	180.00	1195.43	206.31	1581.74

	Unit	Hours	Hours £	Mat'ls £	O & P £	Total £

Furniture and fittings

Take down damaged fittings and
replace (material prices not included
due to the wide variation in the
quality and costs of the fittings)

wall units 2000mm high						
300 × 600mm	nr	1.00	12.00	0.00	1.80	13.80
300 × 1000mm	nr	1.10	13.20	0.00	1.98	15.18
300 × 1200mm	nr	1.20	14.40	0.00	2.16	16.56

base units 750mm high						
600 × 900mm	nr	1.00	12.00	0.00	1.80	13.80
600 × 1000mm	nr	1.00	12.00	0.00	1.80	13.80
600 × 1200mm	nr	1.10	13.20	0.00	1.98	15.18
900 × 900mm	nr	1.10	13.20	0.00	1.98	15.18
900 × 1000mm	nr	1.20	14.40	0.00	2.16	16.56
900 × 1200mm	nr	1.20	14.40	0.00	2.16	16.56

sink units 750mm high						
600 × 900mm	nr	1.40	16.80	0.00	2.52	19.32
600 × 1000mm	nr	1.40	16.80	0.00	2.52	19.32
600 × 1200mm	nr	1.40	16.80	0.00	2.52	19.32

worktops						
600 × 900mm	nr	0.80	9.60	0.00	1.44	11.04
600 × 1000mm	nr	0.90	10.80	0.00	1.62	12.42
600 × 1200mm	nr	1.00	12.00	0.00	1.80	13.80

Remove smoke damaged furniture
and place in skip

easy chair	nr	0.15	1.80	0.00	0.27	2.07
settee	nr	0.18	2.16	0.00	0.32	2.48
single bed and mattress	nr	0.20	2.40	0.00	0.36	2.76
double bed and mattress	nr	0.22	2.64	0.00	0.40	3.04
set of six dining chairs	nr	0.12	1.44	0.00	0.22	1.66
dining table	nr	0.18	2.16	0.00	0.32	2.48
sideboard	nr	0.20	2.40	0.00	0.36	2.76

	Unit	Hours	Hours £	Mat'ls £	O & P £	Total £
Remove smoke damaged furniture (cont'd)						
wardrobe	nr	0.15	1.80	0.00	0.27	2.07
chest of drawers	nr	0.20	2.40	0.00	0.36	2.76
carpet	nr	0.25	3.00	0.00	0.45	3.45
TV and hi-fi equipment	nr	0.25	3.00	0.00	0.45	3.45
Plumbing and heating work						
Take out damaged sanitary fittings and associated pipework and replace						
lavatory basin	nr	3.30	47.85	122.19	25.51	195.55
low level WC	nr	3.40	49.30	311.59	54.13	415.02
shower cubicle	nr	4.20	60.90	847.12	136.20	1044.22
sink with single drainer	nr	2.30	33.35	171.63	30.75	235.73
sink with double drainer	nr	2.90	42.05	216.37	38.76	297.18
bath, cast iron	nr	4.40	63.80	441.04	75.73	580.57
bath, acrylic	nr	4.20	60.90	265.29	48.93	375.12
bidet	nr	3.50	50.75	364.89	62.35	477.99
Take out damaged cisterns and cylinders and associated pipework and replace						
polyethylene cold water cisterns						
68 litres	nr	2.00	29.00	83.75	16.91	129.66
86 litres	nr	2.00	29.00	93.89	18.43	141.32
191 litres	nr	2.50	36.25	125.96	24.33	186.54
327 litres	nr	3.80	55.10	154.34	31.42	240.86
copper cylinders, indirect pattern						
114 litres	nr	3.50	50.75	120.27	25.65	196.67
117 litres	nr	3.50	50.75	127.45	26.73	204.93
140 litres	nr	4.00	58.00	139.76	29.66	227.42
162 litres	nr	4.10	59.45	170.87	34.55	264.87

	Unit	Hours	Hours £	Mat'ls £	O & P £	Total £
copper cylinders, direct pattern						
116 litres	nr	3.80	55.10	117.72	25.92	198.74
120 litres	nr	3.80	55.10	132.07	28.08	215.25
144 litres	nr	4.20	60.90	140.77	30.25	231.92
166 litres	nr	4.30	62.35	158.26	33.09	253.70
Disconnect damaged central heating boiler and associated pipe work and replace						
floor-mounted gas boiler						
30,000 Btu	nr	9.00	130.50	556.82	103.10	790.42
40,000 Btu	nr	9.00	130.50	583.88	107.16	821.54
50,000 Btu	nr	9.20	133.40	655.78	118.38	907.56
60,000 Btu	nr	9.20	133.40	737.80	130.68	1001.88
wall-mounted gas boiler						
30,000 Btu	nr	9.00	130.50	618.73	112.38	861.61
40,000 Btu	nr	9.00	130.50	666.47	119.55	916.52
50,000 Btu	nr	9.20	133.40	724.99	128.76	987.15
60,000 Btu	nr	9.20	133.40	780.39	137.07	1050.86
floor-mounted oil-fired boiler						
52,000 Btu	nr	9.20	133.40	1292.00	213.81	1639.21
70,000 Btu	nr	9.40	136.30	1339.44	221.36	1697.10
Disconnect damaged radiator and associated pipework and and valves and replace						
single panel, 450mm high						
length, 1000mm	nr	1.65	23.93	67.54	13.72	105.18
length, 1600mm	nr	1.85	26.83	94.55	18.21	139.58
length, 2000mm	nr	2.10	30.45	111.75	21.33	163.53
single panel, 600mm high						
length, 1000mm	nr	1.85	26.83	82.67	16.42	125.92
length, 1600mm	nr	2.05	29.73	116.40	21.92	168.04
length, 2000mm	nr	2.30	33.35	139.96	26.00	199.31

	Unit	Hours	Hours £	Mat'ls £	O & P £	Total £
double panel, 450mm high						
length, 1000mm	nr	1.85	26.83	96.07	18.43	141.33
length, 1400mm	nr	1.35	19.58	130.04	22.44	172.06
length, 2000mm	nr	1.45	21.03	165.19	27.93	214.15
double panel, 600mm high						
length, 1000mm	nr	2.05	29.73	93.52	18.49	141.73
length, 1400mm	nr	2.25	32.63	132.28	24.74	189.64
length, 2000mm	nr	2.45	35.53	225.86	39.21	300.59

Electrics

Disconnect power supply, remove
damaged electric fittings and
replace

	Unit	Hours	Hours £	Mat'ls £	O & P £	Total £
single power point	nr	0.60	9.30	6.56	2.38	18.24
double power point	nr	0.60	9.30	7.21	2.48	18.99
light switch	nr	0.60	9.30	5.47	2.22	16.99
light point	nr	0.60	9.30	6.12	2.31	17.73
wall light	nr	0.90	13.95	21.52	5.32	40.79
cooker control unit	nr	1.50	23.25	17.62	6.13	47.00

Painting and decorating

Burn off damaged woodwork and
leave ready to receive new paintwork

	Unit	Hours	Hours £	Mat'ls £	O & P £	Total £
general surfaces	m2	1.10	13.20	0.00	1.98	15.18
general surfaces up to 300mm wide	m	0.30	3.60	0.00	0.54	4.14

Burn off damaged metalwork and
leave ready to receive new paintwork

	Unit	Hours	Hours £	Mat'ls £	O & P £	Total £
general surfaces	m2	1.10	13.20	0.00	1.98	15.18
general surfaces up to 300mm wide	m	0.30	3.60	0.00	0.54	4.14

	Unit	Hours	Hours £	Mat'ls £	O & P £	Total £
Wash down existing plastered surfaces, stop cracks and rub down and leave ready to receive new paintwork						
walls	m2	0.14	1.68	0.00	0.25	1.93
ceilings	m2	0.05	0.60	0.00	0.09	0.69
Apply one mist coat and two coats emulsion paint to plastered surfaces						
walls	m2	0.30	3.60	0.63	0.63	4.86
ceilings	m2	0.32	3.84	0.63	0.67	5.14
Apply two undercoats and one coat gloss paint to plastered surfaces						
walls	m2	0.30	3.60	1.64	0.79	6.03
ceilings	m2	0.36	4.32	1.64	0.89	6.85
Prepare, size, apply adhesive, supply and hang paper to plastered walls						
lining paper						
£1.50 per roll	m2	0.30	3.60	0.38	0.60	4.58
£2.00 per roll	m2	0.30	3.60	0.51	0.62	4.73
£2.50 per roll	m2	0.30	3.60	0.63	0.63	4.86
washable paper						
£2.50 per roll	m2	0.30	3.60	0.63	0.63	4.86
£4.00 per roll	m2	0.30	3.60	1.00	0.69	5.29
£5.00 per roll	m2	0.30	3.60	1.26	0.73	5.59
vinyl paper						
£4.00 per roll	m2	0.30	3.60	1.00	0.69	5.29
£5.00 per roll	m2	0.30	3.60	1.26	0.73	5.59
£6.00 per roll	m2	0.30	3.60	1.50	0.77	5.87

	Unit	Hours	Hours £	Plant £	O & P £	Total £

FLOOD DAMAGE

Pumping

	Unit	Hours	Hours £	Plant £	O & P £	Total £
Instal pump and hoses in position	Item	1.00	12.00	0.00	1.80	13.80
Hire diaphragm pump and hoses						
50mm	hour	1.00	0.00	3.21	0.48	3.69
75mm	hour	1.00	0.00	3.49	0.52	4.01
Hire portable dryer, 350W	hour	1.00	0.00	2.03	0.30	2.33

	Unit	Hours	Hours £	Mat'ls £	O & P £	Total £
Clean up and remove debris from basement and ground floor	m2	0.20	1.96	0.00	0.29	2.25
Hack off defective wall plaster and replace	m2	0.20	2.40	0.00	0.36	2.76
Pull down plasterboard and skim ceilings and renew	m2	1.56	18.72	3.28	3.30	25.30
Hack off damaged wall tiles and renew	m2	1.45	17.40	22.68	6.01	46.09
Take off damaged skirting and replace						
19 × 100mm softwood	m	0.30	3.60	2.32	0.89	6.81
25 × 150mm hardwood	m	0.38	4.56	9.88	2.17	16.61
Take up damaged flooring and renew						
19mm tongued and grooved softwood flooring	m2	1.14	13.68	10.05	3.56	27.29
chipboard flooring	m2	0.80	9.60	5.22	2.22	17.04
18mm tongued and grooved chipboard flooring	m2	1.00	12.00	5.62	2.64	20.26

	Unit	Hours	Hours £	Mat'ls £	O & P £	Total £
Electrics						
Disconnect power supply, remove damaged electric fittings and replace						
single power point	nr	0.60	9.30	6.56	2.38	18.24
double power point	nr	0.60	9.30	7.21	2.48	18.99
light switch	nr	0.60	9.30	5.47	2.22	16.99
cooker control unit	nr	1.50	23.25	17.62	6.13	47.00
Painting and decorating						
Burn off damaged woodwork and leave ready to receive new paintwork						
general surfaces	m2	1.10	13.20	0.00	1.98	15.18
general surfaces up to 300mm wide	m	0.30	3.60	0.00	0.54	4.14
Burn off damaged metalwork and leave ready to receive new paintwork						
general surfaces	m2	1.10	13.20	0.00	1.98	15.18
general surfaces up to 300mm wide	m	0.30	3.60	0.00	0.54	4.14
Wash down existing plastered surfaces, stop cracks and rub down and leave ready to receive new paintwork						
walls	m2	0.05	0.60	0.00	0.09	0.69
ceilings	m2	0.07	0.84	0.00	0.13	0.97
Apply two undercoats and one coat gloss paint to plastered surfaces						
walls	m2	0.30	3.60	1.64	0.79	6.03
ceilings	m2	0.36	4.32	1.64	0.89	6.85

	Unit	Hours	Hours £	Mat'ls £	O & P £	Total £

GALE DAMAGE

Hoardings and screens

	Unit	Hours	Hours £	Mat'ls £	O & P £	Total £
Temporary screens and hoardings 2m high consisting of 22mm thick exterior quality plywood fixed to 100 × 50mm posts and rails	m	2.60	31.20	31.15	9.35	71.70
Tarpaulin sheeting size 5 × 4m fixed in position	day	0.00	0.00	3.81	0.57	4.38
Plywood sheeting blocking up window or door opening	m2	1.00	12.00	15.58	4.14	31.72

Shoring

	Unit	Hours	Hours £	Mat'ls £	O & P £	Total £
Timber dead shores consisting of 200 × 200mm shores, 200 × 50mm plates and 200 × 50mm braces at centres of						
2m	m2	3.80	45.60	72.24	17.68	135.52
3m	m2	3.00	36.00	58.00	14.10	108.10
4m	m2	2.20	26.40	41.68	10.21	78.29
Timber raking and flying shores consisting of 200 × 200mm shores, 250 × 50mm plates and 200 × 50mm braces to gable end of two-storey house	Item	72.00	864.00	520.40	207.66	1592.06

Roof repairs

	Unit	Hours	Hours £	Mat'ls £	O & P £	Total £
Take up roof coverings from pitched roof						
tiles	m2	0.80	7.84	0.00	1.18	9.02
slates	m2	0.80	7.84	0.00	1.18	9.02
timber boarding	m2	1.00	9.80	0.00	1.47	11.27
flat sheeting	m2	0.30	2.94	0.00	0.44	3.38
corrugated sheeting	m2	0.30	2.94	0.00	0.44	3.38

	Unit	Hours	Hours £	Mat'ls £	O & P £	Total £
Take up roof coverings from flat roof						
bituminous felt	m2	0.25	2.45	0.00	0.37	2.82
metal sheeting	m2	0.30	2.94	0.00	0.44	3.38
woodwool slabs	m2	0.50	4.90	0.00	0.74	5.64
firrings	m2	0.20	1.96	0.00	0.29	2.25
Take up roof coverings from pitched roof, carefully lay aside for reuse						
tiles	m2	1.10	10.78	0.00	1.62	12.40
slates	m2	1.10	10.78	0.00	1.62	12.40
metal sheeting	m2	0.50	4.90	0.00	0.74	5.64
flat sheeting	m2	0.60	5.88	0.00	0.88	6.76
corrugated sheeting	m2	0.60	5.88	0.00	0.88	6.76
Take up roof coverings from flat roof, carefully lay aside for reuse						
metal sheeting	m2	0.60	5.88	0.00	0.88	6.76
woodwool slabs	m2	0.80	7.84	0.00	1.18	9.02
Inspect roof battens, refix loose and replace with new, size 38 × 25mm						
25% of area						
250mm centres	m2	0.14	1.68	0.42	0.32	2.42
450mm centres	m2	0.12	1.44	0.37	0.27	2.08
600mm centres	m2	0.10	1.20	0.32	0.23	1.75
75% of area						
250mm centres	m2	0.36	4.32	1.07	0.81	6.20
450mm centres	m2	0.26	3.12	0.90	0.60	4.62
600mm centres	m2	0.22	2.64	0.74	0.51	3.89
100% of area						
250mm centres	m2	0.44	5.28	1.39	1.00	7.67
450mm centres	m2	0.32	3.84	1.17	0.75	5.76
600mm centres	m2	0.38	4.56	0.96	0.83	6.35

	Unit	Hours	Hours £	Mat'ls £	O & P £	Total £
Remove single slipped slate and refix	nr	1.00	12.00	0.96	1.94	14.90
Remove single broken slate, renew with new Welsh blue slate						
405 × 255mm	nr	1.20	14.40	1.39	2.37	18.16
510 × 255mm	nr	1.20	14.40	2.67	2.56	19.63
610 × 305mm	nr	1.20	14.40	5.47	2.98	22.85
Remove slates in area approximately 1m2 and replace with Welsh slates previously laid aside	nr	1.60	19.20	0.00	2.88	22.08
Remove slates in area approximately 1m2 and replace with Welsh blue slates previously laid aside						
405 × 255mm	nr	1.80	21.60	58.16	11.96	91.72
510 × 255mm	nr	1.70	20.40	61.94	12.35	94.69
610 × 305mm	nr	1.60	19.20	64.45	12.55	96.20
Remove double course at eaves and fix new Welsh blue slates						
405 × 255mm	m	0.70	8.40	26.27	5.20	39.87
510 × 255mm	m	0.70	8.40	32.24	6.10	46.74
610 × 305mm	m	0.70	8.40	37.31	6.86	52.57
Remove single slipped tile and refix	nr	0.30	3.60	0.00	0.54	4.14
Remove single broken tile and renew						
Marley plain tile	nr	0.30	3.60	0.70	0.65	4.95
Marley Ludlow Plus tile	nr	0.30	3.60	0.89	0.67	5.16
Marley Modern tile	nr	0.30	3.60	1.44	0.76	5.80
Redland Norfolk tile	nr	0.30	3.60	1.21	0.72	5.53
Remove tiles in area approximately 1m2 and replace with tiles previously laid aside						
Marley plain tile	nr	1.80	21.60	29.84	7.72	59.16
Marley Ludlow Plus tile	nr	1.20	14.40	14.50	4.34	33.24
Marley Modern tile	nr	1.10	13.20	15.26	4.27	32.73

	Unit	Hours	Hours £	Mat'ls £	O & P £	Total £
Take off defective ridge capping and refix including pointing in mortar	m	1.10	13.20	1.55	2.21	16.96

Chimney stack

	Unit	Hours	Hours £	Mat'ls £	O & P £	Total £
Erect and take down chimney scaffold	item	4.00	48.00	0.00	7.20	55.20
Hire chimney scaffold and platform	week	0.00	0.00	100.58	15.09	115.67
Take down existing chimney stack to below roof level and remove debris						
single stack	m	2.50	30.00	1.41	4.71	36.12
double stack	m	3.50	42.00	1.41	6.51	49.92
Chimney stack in facing brick £450.00 per thousand in gauged mortar						
single stack	m	3.80	45.60	45.54	13.67	104.81
double stack	m	5.20	62.40	94.72	23.57	180.69
Terra cotta chimney pot bedded and flaunched in gauged mortar 185mm diameter × 300mm high	nr	1.25	15.00	25.91	6.14	47.05
185mm diameter × 600mm high	nr	1.80	21.60	44.52	9.92	76.04

External work

	Unit	Hours	Hours £	Mat'ls £	O & P £	Total £
Remove blown-down trees, trunk girth 1m above ground level						
600 to 1500mm	nr	22.00	264.00	0.00	39.60	303.60
1500 to 3000mm	nr	38.00	456.00	0.00	68.40	524.40

	Unit	Hours	Hours £	Mat'ls £	O & P £	Total £

THEFT DAMAGE

Window and door repairs

	Unit	Hours	Hours £	Mat'ls £	O & P £	Total £
Hack out glass and remove	m2	0.45	5.40	0.00	0.81	6.21
Clean rebates, remove sprigs or clips and prepare for reglazing	m	0.20	2.40	0.00	0.36	2.76

Reglaze existing softwood windows in clear float glass with putty and sprigs

under 0.15m2, thickness

	Unit	Hours	Hours £	Mat'ls £	O & P £	Total £
4mm	m2	0.90	10.80	28.15	5.84	44.79
5mm	m2	0.90	10.80	31.93	6.41	49.14
6mm	m2	1.00	12.00	33.46	6.82	52.28
10mm	m2	1.05	12.60	58.43	10.65	81.68

over 0.15m2, thickness

	Unit	Hours	Hours £	Mat'ls £	O & P £	Total £
4mm	m2	0.60	7.20	26.83	5.10	39.13
5mm	m2	0.60	7.20	30.89	5.71	43.80
6mm	m2	0.65	7.80	32.68	6.07	46.55
10mm	m2	0.70	8.40	56.03	9.66	74.09

Reglaze existing softwood windows in clear float glass with pinned beads

under 0.15m2, thickness

	Unit	Hours	Hours £	Mat'ls £	O & P £	Total £
4mm	m2	1.10	13.20	28.15	6.20	47.55
5mm	m2	1.10	13.20	31.93	6.77	51.90
6mm	m2	1.20	14.40	33.46	7.18	55.04
10mm	m2	1.25	15.00	58.43	11.01	84.44

over 0.15m2, thickness

	Unit	Hours	Hours £	Mat'ls £	O & P £	Total £
4mm	m2	0.80	9.60	26.83	5.46	41.89
5mm	m2	0.80	9.60	30.89	6.07	46.56
6mm	m2	0.90	10.80	32.68	6.52	50.00
10mm	m2	0.95	11.40	56.03	10.11	77.54

	Unit	Hours	Hours £	Mat'ls £	O & P £	Total £

**Reglaze existing metal windows
in clear float glass with clips and
putty**
under 0.15m2, thickness

	Unit	Hours	Hours £	Mat'ls £	O & P £	Total £
4mm	m2	0.95	11.40	31.68	6.46	49.54
5mm	m2	0.95	11.40	34.73	6.92	53.05
6mm	m2	1.05	12.60	37.35	7.49	57.44
10mm	m2	1.10	13.20	60.31	11.03	84.54

over 0.15m2, thickness

	Unit	Hours	Hours £	Mat'ls £	O & P £	Total £
4mm	m2	0.65	7.80	30.61	5.76	44.17
5mm	m2	0.65	7.80	33.45	6.19	47.44
6mm	m2	0.70	8.40	35.97	6.66	51.03
10mm	m2	0.75	9.00	59.18	10.23	78.41

**Reglaze existing metal windows
in clear float glass with screwed
metal beads**
under 0.15m2, thickness

	Unit	Hours	Hours £	Mat'ls £	O & P £	Total £
4mm	m2	1.30	15.60	35.85	7.72	59.17
5mm	m2	1.30	15.60	38.72	8.15	62.47
6mm	m2	1.40	16.80	41.66	8.77	67.23
10mm	m2	1.45	17.40	65.62	12.45	95.47

over 0.15m2, thickness

	Unit	Hours	Hours £	Mat'ls £	O & P £	Total £
4mm	m2	1.00	12.00	34.95	7.04	53.99
5mm	m2	1.00	12.00	38.12	7.52	57.64
6mm	m2	1.05	12.60	40.02	7.89	60.51
10mm	m2	1.10	13.20	64.99	11.73	89.92

**Reglaze existing softwood windows
in white patterned glass with putty
and sprigs**
under 0.15m2, thickness

	Unit	Hours	Hours £	Mat'ls £	O & P £	Total £
4mm	m2	0.90	10.80	31.39	6.33	48.52
6mm	m2	1.00	12.00	36.81	7.32	56.13

over 0.15m2, thickness

	Unit	Hours	Hours £	Mat'ls £	O & P £	Total £
4mm	m2	0.60	7.20	31.39	5.79	44.38
6mm	m2	0.65	7.80	36.81	6.69	51.30

	Unit	Hours	Hours £	Mat'ls £	O & P £	Total £
Reglaze existing softwood windows in white patterned glass with pinned beads						
under 0.15m2, thickness						
4mm	m2	1.10	13.20	31.39	6.69	51.28
6mm	m2	1.20	14.40	36.81	7.68	58.89
over 0.15m2, thickness						
4mm	m2	0.80	9.60	31.39	6.15	47.14
6mm	m2	0.85	10.20	36.81	7.05	54.06
Reglaze existing softwood windows in white patterned glass with screwed beads						
under 0.15m2, thickness						
4mm	m2	1.30	15.60	33.43	7.35	56.38
6mm	m2	1.40	16.80	38.91	8.36	64.07
over 0.15m2, thickness						
4mm	m2	1.00	12.00	33.43	6.81	52.24
6mm	m2	1.50	18.00	33.43	7.71	59.14
Reglaze existing metal windows in white patterned glass with putty						
under 0.15m2, thickness						
4mm	m2	0.90	10.80	31.39	6.33	48.52
6mm	m2	1.00	12.00	36.81	7.32	56.13
over 0.15m2, thickness						
4mm	m2	0.60	7.20	31.39	5.79	44.38
6mm	m2	0.65	7.80	36.81	6.69	51.30
Reglaze existing metal windows in white patterned glass with metal clips and putty						
under 0.15m2, thickness						
4mm	m2	0.95	11.40	33.43	6.72	51.55
6mm	m2	1.05	12.60	33.43	6.90	52.93

	Unit	Hours	Hours £	Mat'ls £	O & P £	Total £
over 0.15m2, thickness						
4mm	m2	0.65	7.80	33.43	6.18	47.41
6mm	m2	0.70	8.40	33.43	6.27	48.10
Reglaze existing metal windows in white patterned glass with screwed beads						
under 0.15m2, thickness						
4mm	m2	1.30	15.60	29.58	6.78	51.96
6mm	m2	1.40	16.80	34.71	7.73	59.24
over 0.15m2, thickness						
4mm	m2	1.00	12.00	29.58	6.24	47.82
6mm	m2	1.05	12.60	34.71	7.10	54.41
Take off and replace damaged ironmongery to softwood windows						
casement fastener	nr	0.20	2.40	3.90	0.95	7.25
casement stay	nr	0.20	2.40	5.93	1.25	9.58
hinges	pair	0.25	3.00	3.54	0.98	7.52
lock	nr	0.60	7.20	12.01	2.88	22.09
Take off and replace damaged ironmongery to hardwood windows						
casement fastener	nr	0.30	3.60	3.90	1.13	8.63
casement stay	nr	0.30	3.60	5.93	1.43	10.96
hinges	pair	0.35	4.20	3.54	1.16	8.90
lock	nr	0.75	9.00	12.01	3.15	24.16
Take off and replace defective ironmongery to PVC-U windows						
casement fastener	nr	0.20	2.40	3.90	0.95	7.25
casement stay	nr	0.20	2.40	5.93	1.25	9.58
hinges	pair	0.25	3.00	3.54	0.98	7.52
lock	nr	0.60	7.20	12.01	2.88	22.09
Take out existing window cill and replace						
softwood	m	0.55	6.60	8.07	2.20	16.87
hardwood	m	0.65	7.80	23.92	4.76	36.48

	Unit	Hours	Hours £	Mat'ls £	O & P £	Total £
Take off existing external door and frame and replace with new						
flush door						
softwood	nr	2.20	26.40	93.49	17.98	137.87
hardwood	nr	2.50	30.00	336.26	54.94	421.20
PVC-U	nr	2.20	26.40	309.56	50.39	386.35
panelled door						
softwood	nr	2.20	26.40	117.82	21.63	165.85
hardwood	nr	2.50	30.00	368.67	59.80	458.47
PVC-U	nr	2.20	26.40	345.97	55.86	428.23
half glazed door						
softwood	nr	2.20	26.40	117.82	21.63	165.85
hardwood	nr	2.50	30.00	368.67	59.80	458.47
PVC-U	nr	2.20	26.40	345.97	55.86	428.23
fully glazed door						
softwood	nr	2.80	33.60	94.52	19.22	147.34
hardwood	nr	3.20	38.40	348.43	58.02	444.85
PVC-U	nr	2.80	33.60	321.54	53.27	408.41
fully glazed patio doors						
softwood	pair	3.80	45.60	212.70	38.75	297.05
hardwood	pair	4.30	51.60	489.40	81.15	622.15
PVC-U	pair	2.80	33.60	451.68	72.79	558.07
Take off and replace damaged ironmongery to softwood doors						
bolts						
barrel	nr	0.20	2.40	5.92	1.25	9.57
flush	nr	0.20	2.40	7.72	1.52	11.64
tower	nr	0.30	3.60	6.77	1.56	11.93
butts						
light	pair	0.20	2.40	3.55	0.89	6.84
medium	pair	0.25	3.00	3.61	0.99	7.60
heavy	pair	0.30	3.60	4.54	1.22	9.36

	Unit	Hours	Hours £	Mat'ls £	O & P £	Total £

**Take off and replace damaged
ironmongery to softwood doors
(cont'd)**

locks
cupboard	nr	0.30	3.60	6.99	1.59	12.18
mortice dead lock	nr	0.85	10.20	12.02	3.33	25.55
rim lock	nr	0.45	5.40	5.46	1.63	12.49
cylinder	nr	1.10	13.20	23.89	5.56	42.65

Take off and replace damaged
ironmongery to hardwood doors

bolts
barrel	nr	0.30	3.60	5.92	1.43	10.95
flush	nr	0.30	3.60	7.72	1.70	13.02
tower	nr	0.40	4.80	6.77	1.74	13.31

butts
light	pair	0.30	3.60	3.55	1.07	8.22
medium	pair	0.35	4.20	3.61	1.17	8.98
heavy	pair	0.40	4.80	4.54	1.40	10.74

locks
cupboard	nr	0.30	3.60	6.99	1.59	12.18
mortice dead lock	nr	0.85	10.20	12.02	3.33	25.55
rim lock	nr	0.45	5.40	5.46	1.63	12.49
cylinder	nr	1.10	13.20	23.89	5.56	42.65

Part Three

APPROXIMATE ESTIMATING

Excavation and filling

Concrete work

Brickwork and blockwork

Masonry

Carpentry and joinery

Finishings

Plumbing and heating

Painting

Wallpapering

Drainage

	Unit	Total £

APPROXIMATE ESTIMATING

The following rates are based upon
the unit rates in Part One of this
book but have been rounded off for
ease of use

Excavation and filling

Excavate for trench including
supporting sides, level and ram
bottom, part return, fill and ram,
part load into skip, width 600mm

	Unit	Total £
by hand, depth		
0.75m	m	25.00
1.00m	m	31.00
1.50m	m	44.00
2.00m	m	55.00
by machine, depth		
0.75m	m	17.00
1.00m	m	21.00
1.50m	m	27.00
2.00m	m	34.00
Filling in layers by hand over 250mm thick		
surplus excavated material	m3	17.00
sand	m3	40.00
hardcore	m3	34.00
Filling in layers by machine over 250mm thick		
surplus excavated material	m3	9.00
sand	m3	34.00
hardcore	m3	36.00

	Unit	Total £

Concrete work

Ready mixed concrete 1:3:6
40mm aggregate in foundations,
size

600 x 225mm	m	17.00
750 x 225mm	m	21.00

Ready mixed concrete 1:2:4
20mm aggregate, in beds including
mesh reinforcement and trowelling
smooth

beds, thickness		
100mm	m2	21.00
150mm	m2	27.00
200mm	m2	34.00

Ready mixed concrete 1:2:4
20mm aggregate, in slabs including
mesh reinforcement and formwork
to soffit

100mm	m2	78.00
150mm	m2	85.00
200mm	m2	92.00

Ready mixed concrete 1:2:4
20mm aggregate, in walls including
mesh reinforcement and formwork
both sides

100mm	m2	85.00
150mm	m2	93.00
200mm	m2	101.00

	Unit	Total £

Brickwork and blockwork

Cavity wall in cement mortar
in foundations including
forming cavity and wall ties
between outer leaf in common
bricks £140 per 1000 and
inner leaf of

common brick £140 per 1000	m2	104.00
blockwork 100mm thick	m2	90.00

Cavity wall in cement mortar
in foundations including
forming cavity and wall ties
between outer leaf in common
bricks £200 per 1000 and
inner leaf of

common brick £200 per 1000	m2	116.00
blockwork 100mm thick	m2	96.00

Cavity wall in cement mortar
in foundations including
forming cavity and wall ties
between outer leaf in engineering
bricks £250 per 1000 and
inner leaf of

engineering bricks £250 per 1000	m2	129.00
blockwork 100mm thick	m2	103.00

Cavity wall in cement mortar
in foundations including forming
cavity and wall ties between outer
leaf in engineering bricks £350 per
1000 and inner leaf of

engineering bricks £350 per 1000	m2	151.00
blockwork 100mm thick	m2	113.00

	Unit	Total £
Cavity wall in gauged mortar including forming cavity and wall ties between outer leaf in facing bricks £250 per 1000 and inner leaf of		
common brick £140 per 1000	m2	78.00
common brick £200 per 1000	m2	131.00
blockwork 100mm thick	m2	111.00
Cavity wall in gauged mortar including forming cavity and wall ties between outer leaf in facing bricks £400 per 1000 and inner leaf of		
common brick £140 per 1000	m2	137.00
common brick £200 per 1000	m2	143.00
blockwork 100mm thick	m2	123.00
Cavity wall in gauged mortar including forming cavity and wall ties between outer leaf in facing bricks £500 per 1000 and inner leaf of		
common brick £140 per 1000	m2	144.00
common brick £200 per 1000	m2	150.00
blockwork 100mm thick	m2	130.00

Masonry

	Unit	Total £
Cavity wall in gauged mortar including forming cavity and wall ties between outer leaf in irregular coursed rubble walling 150mm wide and inner leaf of		
common brick £140 per 1000	m2	134.00
common brick £200 per 1000	m2	140.00

	Unit	Total £
Cavity wall in gauged mortar including forming cavity and wall ties between outer leaf in coursed rubble walling 150mm wide and inner leaf of		
common brick £140 per 1000	m2	140.00
common brick £200 per 1000	m2	146.00

Carpentry and joinery

	Unit	Total £
19mm thick butt-jointed flooring fixed to softwood joists size		
50 x 100mm	m2	25.00
50 x 125mm	m2	26.00
50 x 150mm	m2	27.00
25mm thick butt-jointed flooring fixed to softwood joists size		
50 x 100mm	m2	29.00
50 x 125mm	m2	30.00
50 x 150mm	m2	31.00
19mm thick tongued and grooved flooring fixed to softwood joists size		
50 x 100mm	m2	29.00
50 x 125mm	m2	30.00
50 x 150mm	m2	31.00
25mm thick tongued and grooved flooring fixed to softwood joists size		
50 x 100mm	m2	31.00
50 x 125mm	m2	32.00
50 x 150mm	m2	33.00

	Unit	Total £
18mm thick chipboard butt-jointed flooring fixed to softwood joists size		
50 x 100mm	m2	19.00
50 x 125mm	m2	21.00
50 x 150mm	m2	22.00
25mm thick chipboard butt-jointed flooring fixed to softwood joists size		
50 x 100mm	m2	22.00
50 x 125mm	m2	23.00
50 x 150mm	m2	24.00
25mm thick chipboard tongued and grooved flooring fixed to softwood joists size		
50 x 100mm	m2	24.00
50 x 125mm	m2	25.00
50 x 150mm	m2	26.00
Standard flush hardboard-faced internal door including lining and stops, architraves and fixing only hardware		
35mm thick		
610 x 1981mm	nr	142.00
686 x 1981mm	nr	145.00
762 x 1981mm	nr	147.00
40mm thick		
610 x 1981mm	nr	146.00
686 x 1981mm	nr	148.00
762 x 1981mm	nr	150.00

	Unit	Total £
Standard flush sapele-faced internal door including lining and stops, architraves and fixing only hardware		
35mm thick		
610 x 1981mm	nr	160.00
686 x 1981mm	nr	162.00
762 x 1981mm	nr	164.00
40mm thick		
610 x 1981mm	nr	165.00
686 x 1981mm	nr	168.00
762 x 1981mm	nr	170.00
Standard flush hardboard-faced half-hour fire-check internal door including lining and stops, architraves and fixing only hardware		
44mm thick		
686 x 1981mm	nr	201.00
762 x 1981mm	nr	205.00
726 x 2040mm	nr	209.00
826 x 2040mm	nr	214.00
Standard flush sapele-faced half-hour fire-check internal door including lining and stops, architraves and fixing only hardware		
44mm thick		
686 x 1981mm	nr	230.00
762 x 1981mm	nr	235.00
726 x 2040mm	nr	241.00
826 x 2040mm	nr	247.00

	Unit	Total £
Standard flush hardboard-faced external door including frame, architraves and fixing only hardware		
44mm thick		
762 x 1981mm	nr	241.00
838 x 1981mm	nr	253.00

Finishings

	Unit	Total £
One coat 'Universal' plaster to walls, plasterboard and skim to ceilings in rooms, storey height 2.4m		
floor area		
9m2	nr	385.00
12m2	nr	477.00
15m2	nr	569.00
18m2	nr	638.00
21m2	nr	713.00
24m2	nr	805.00
27m2	nr	894.00
30m2	nr	943.00

Plumbing and heating

PVC-U rainwater goods

	Unit	Total £
one-storey gable end		
terraced	nr	240.00
semi-detached	nr	288.00
detached	nr	336.00
two-storey gable end		
terraced	nr	300.00
semi-detached	nr	342.00
detached	nr	372.00

	Unit	Total £
three-storey gable end		
terraced	nr	378.00
semi-detached	nr	402.00
detached	nr	444.00
one-storey hipped end		
terraced	nr	240.00
semi-detached	nr	360.00
detached	nr	456.00
two-storey hipped end		
terraced	nr	336.00
semi-detached	nr	432.00
detached	nr	528.00
three-storey hipped end		
terraced	nr	396.00
semi-detached	nr	504.00
detached	nr	612.00
Cast iron rainwater goods		
one-storey gable end		
terraced	nr	636.00
semi-detached	nr	708.00
detached	nr	750.00
two-storey gable end		
terraced	nr	828.00
semi-detached	nr	888.00
detached	nr	936.00
three-storey gable end		
terraced	nr	972.00
semi-detached	nr	1116.00
detached	nr	1164.00

	Unit	Total £
one-storey hipped end		
terraced	nr	612.00
semi-detached	nr	816.00
detached	nr	1056.00
two-storey hipped end		
terraced	nr	792.00
semi-detached	nr	1088.00
detached	nr	1272.00
three-storey hipped end		
terraced	nr	936.00
semi-detached	nr	1212.00
detached	nr	1440.00

Central heating

	Unit	Total £
one-storey house overall size		
(5 radiators)	nr	2,160.00
(6 radiators)	nr	2,340.00
(7 radiators)	nr	2,490.00
two-storey house overall size		
(8 radiators)	nr	2,940.00
(9 radiators)	nr	3,180.00
(10 radiators)	nr	3,540.00
three-storey house overall size		
(12 radiators)	nr	4,920.00
(14 radiators)	nr	5,220.00
(15 radiators)	nr	5,400.00

Painting

One mist coat and two coats of emulsion paint to walls and ceilings in rooms, storey height 2.4m

	Unit	Total £
floor area		
9m2	nr	250.00

	Unit	Total £
12m2	nr	296.00
15m2	nr	347.00
18m2	nr	381.00
21m2	nr	416.00
24m2	nr	467.00
27m2	nr	495.00
30m2	nr	535.00

Wallpapering

Prepare and hang wallpaper £5 per
roll to walls in rooms, storey
height 2.4m

floor area		
9m2	nr	160.00
12m2	nr	188.00
15m2	nr	216.00
18m2	nr	228.00
21m2	nr	245.00
24m2	nr	267.00
27m2	nr	285.00
30m2	nr	296.00

Prepare and hang wallpaper £7 per
roll to walls in rooms, storey
height 2.4m

floor area		
9m2	nr	176.00
12m2	nr	205.00
15m2	nr	239.00
18m2	nr	250.00
21m2	nr	267.00
24m2	nr	296.00
27m2	nr	313.00
30m2	nr	324.00

	Unit	Total £
Prepare and hang wallpaper £9 per roll to walls in rooms, storey height 2.4m		
floor area		
9m2	nr	194.00
12m2	nr	228.00
15m2	nr	256.00
18m2	nr	273.00
21m2	nr	290.00
24m2	nr	324.00
27m2	nr	342.00
30m2	nr	360.00

Drainage

Excavate trench by hand, lay 100mm diameter 'Hepsleve' pipe, granular bed and surround, trench depth

	Unit	Total £
0.50m	m	35.00
0.75m	m	43.00
1.00m	m	49.00
1.25m	m	64.00
1.50m	m	75.00
1.75m	m	85.00
2.00m	m	99.00
2.25m	m	109.00
2.50m	m	123.00
2.75m	m	133.00
3.00m	m	150.00

Excavate trench by hand, lay 150mm diameter 'Hepsleve' pipe, granular bed and surround, trench depth

	Unit	Total £
0.50m	m	48.00
0.75m	m	56.00

	Unit	Total £
1.00m	m	62.00
1.25m	m	77.00
1.50m	m	88.00
1.75m	m	98.00
2.00m	m	111.00
2.25m	m	121.00
2.50m	m	135.00
2.75m	m	146.00
3.00m	m	162.00

Excavate trench by machine, lay
100mm diameter 'Hepsleve' pipe,
granular bed and surround, trench
depth

0.50m	m	15.00
0.75m	m	30.00
1.00m	m	35.00
1.25m	m	45.00
1.50m	m	48.00
1.75m	m	54.00
2.00m	m	62.00
2.25m	m	70.00
2.50m	m	78.00
2.75m	m	87.00
3.00m	m	94.00

Excavate trench by machine, lay
100mm diameter 'Hepsleve' pipe,
granular bed and surround, trench
depth

0.50m	m	40.00
0.75m	m	43.00
1.00m	m	48.00
1.25m	m	58.00
1.50m	m	62.00
1.75m	m	67.00
2.00m	m	74.00
2.25m	m	84.00

	Unit	Total £
2.50m	m	92.00
2.75m	m	100.00
3.00m	m	107.00

Manhole complete including hand
excavation, concrete base and
benching, engineering brickwork,
channels and cast iron cover,
depth

1.00m	nr	480.00
1.50m	nr	650.00
2.00m	nr	840.00

Manhole complete including machine
excavation, concrete base and
benching, engineering brickwork,
channels and cast iron cover,
depth

1.00m	nr	465.00
1.50m	nr	600.00
2.00m	nr	730.00

Part Four

PLANT AND TOOL HIRE

Tool and equipment hire

Access and site equipment

Lifting and moving

Compaction

Breaking and demolition

Power tools

Welding and generators

Pumping equipment

	24 hours	Additional 24 hours	Week
	£	£	£

TOOL AND EQUIPMENT HIRE

These selected rates are based on
average hire charges made by
hire firms in UK. Check your
local dealer for more information.
These prices exclude VAT.

CONCRETE AND CUTTING EQUIPMENT

Concrete mixers

	24 hours	Additional 24 hours	Week
Petrol, with stand	14.00	7.00	24.00
Electric, with stand	13.00	6.00	26.00
Bulk mixer	28.00	14.00	56.00

Vibrating pokers

Pokers			
petrol	42.00	21.00	84.00
electric	34.00	17.00	68.00
air poker, 50mm	28.00	14.00	56.00
air poker, 75mm	30.00	15.00	60.00

Power floats

Floats			
power float, petrol	38.00	19.00	76.00

Vibrating screeds

Screed units			
with 5m beam	54.00	27.00	108.00
with 3.25 - 5.20 beam	58.00	29.00	116.00
roller screed unit	104.00	52.00	208.00

	24 hours	Additional 24 hours	Week
	£	£	£
Floor preparation units			
Floor saw, petrol			
350mm	54.00	27.00	108.00
450mm	68.00	34.00	136.00
Scabbler, hand held	22.00	11.00	44.00
Diamond concrete planer	50.00	25.00	100.00
Air needle gun	22.00	11.00	44.00
Disc cutters			
Cutters			
electric, 300mm	22.00	11.00	44.00
two stroke, 300mm	24.00	12.00	48.00
two stroke, 350mm	28.00	14.00	56.00
electric wall chasers	36.00	18.00	72.00
Block and slab splitters			
Splitters			
clay	50.00	25.00	100.00
block	24.00	12.00	48.00
slab	38.00	19.00	76.00
Saws			
Electric, 150mm	36.00	18.00	72.00
Electric, 300mm	40.00	20.00	80.00

ACCESS AND SITE EQUIPMENT

Ladders

	24 hours	Additional 24 hours	Week
Double ladder, alloy			
4m	14.00	7.00	28.00
6m	16.00	8.00	32.00
9m	18.00	9.00	36.00

	24 hours	Additional 24 hours	Week
	£	£	£
Triple ladder, alloy			
9m	16.00	8.00	32.00
Roof ladder			
5m	16.00	8.00	32.00
Rope operated			
11m	34.00	17.00	68.00
13m	40.00	20.00	80.00
16m	48.00	24.00	96.00

Props

	24 hours	Additional 24 hours	Week
Shoring props			
type 0	0.00	0.00	4.00
type 1	0.00	0.00	4.00
type 2	0.00	0.00	4.00
type 3	0.00	0.00	4.00
type 4	0.00	0.00	4.00

Rubbish chutes

	24 hours	Additional 24 hours	Week
Chutes			
1m section	6.00	3.00	12.00
funnel	6.00	3.00	12.00
frame	6.00	3.00	12.00

Trestles and staging

	24 hours	Additional 24 hours	Week
Staging			
2.4m	14.00	7.00	28.00
3.6m	20.00	10.00	40.00
4.8m	22.00	11.00	44.00
Painters' trestle			
1.8m	10.00	5.00	20.00

	24 hours £	Additional 24 hours £	Week £
Alloy towers			
Single width, height			
2.30m	32.00	16.00	64.00
3.20m	38.00	19.00	76.00
3.73m	40.00	20.00	80.00
4.20m	44.00	22.00	88.00
5.20m	58.00	29.00	116.00
6.52m	62.00	31.00	122.00
7.45m	70.00	35.00	140.00
7.91m	76.00	38.00	152.00
8.20m	84.00	42.00	168.00
9.30m	88.00	44.00	176.00
10.20m	106.00	53.00	212.00
Full width, height			
2.34m	38.00	19.00	76.00
3.27m	42.00	21.00	84.00
3.73m	44.00	22.00	88.00
4.66m	54.00	26.00	104.00
5.59m	56.00	28.00	112.00
6.52m	60.00	30.00	120.00
7.45m	66.00	33.00	132.00
7.91m	70.00	35.00	140.00
8.84m	76.00	38.00	156.00
10.23m	86.00	43.00	172.00
11.16m	94.00	47.00	188.00
12.55m	108.00	54.00	216.00
13.94m	120.00	60.00	240.00
14.87m	126.00	63.00	252.00
15.80m	130.00	65.00	260.00

LIFTING AND MOVING

	24 hours £	Additional 24 hours £	Week £
Sack truck	8.00	4.00	16.00
Pallet truck	28.00	14.00	56.00
Plasterboard jack lift	16.00	8.00	32.00
Stair climber	14.00	7.00	28.00
Rubble truck	20.00	10.00	40.00

	24 hours	Additional 24 hours	Week
	£	£	£

COMPACTION

Plate compactors

Compactors			
350 x 550mm, petrol	54.00	27.00	108.00
700 x 320mm, diesel	58.00	29.00	116.00

Vibrating rollers

Mini			
Large	54.00	27.00	108.00
	64.00	32.00	128.00

BREAKING AND DEMOLITION

Breakers

Hydraulic			
diesel	70.00	35.00	140.00
petrol	76.00	38.00	152.00
medium duty, electric	22.00	11.00	44.00
heavy duty, electric	44.00	22.00	88.00
Air breaker, medium	32.00	16.00	64.00
Air breaker, heavy	34.00	17.00	68.00

POWER TOOLS

Drills

Cordless drill	16.00	8.00	32.00
Cordless impact	20.00	10.00	40.00
Two speed impact	10.00	5.00	20.00
Rotary drill, 16mm	24.00	12.00	48.00
Rotary drill, 20mm	26.00	13.00	52.00
Right angle drill	20.00	10.00	40.00
Combi hammer			
light duty	16.00	8.00	32.00
meduim duty	22.00	11.00	46.00
heavy duty	24.00	12.00	48.00

	24 hours	Additional 24 hours	Week
	£	£	£

Grinders

Angle grinder			
100mm	14.00	7.00	28.00
125mm	14.00	7.00	28.00
230mm	16.00	8.00	32.00
300mm	26.00	13.00	52.00

Saws

Reciprocating saw			
standard	22.00	11.00	44.00
heavy duty	24.00	12.00	48.00
Circular saw			
150mm	16.00	8.00	32.00
230mm	18.00	9.00	36.00
Door trimmer	32.00	16.00	64.00

Woodworking

Plane, 3.25in	16.00	8.00	32.00
Router	16.00	8.00	32.00
Worktop jig	14.00	7.00	28.00

Fixing equipment

Cordless nailing gun	26.00	13.00	52.00
Cartridge hammer	22.00	11.00	44.00
Electric screwdriver	14.00	7.00	28.00

Sanders

Floor	34.00	17.00	68.00
Floor edger	26.00	13.00	52.00
Orbital	18.00	9.00	36.00

	24 hours	Additional 24 hours	Week
	£	£	£

WELDING AND GENERATORS

Generating

Generators
petrol, 2KVA	22.00	11.00	44.00
petrol, 3KVA	32.00	16.00	64.00
silenced, 8KVA	104.00	52.00	208.00
silenced, 15KVA	118.00	59.00	236.00

PUMPING EQUIPMENT

Pumps

Submersible
25mm	16.00	8.00	32.00
50mm	32.00	16.00	64.00
Puddle pump, electric	34.00	17.00	68.00
Diaphragm pump, 50mm	54.00	27.00	108.00

Part Five

GENERAL CONSTRUCTION DATA

GENERAL CONSTRUCTION DATA

The metric system

Linear

1 centimetre (cm)	=	10 millimetres (mm)
1 decimetre (dm)	=	10 centimetres (cm)
1 metre (m)	=	10 decimetres (dm)
1 kilometre (km)	=	1000 metres (m)

Area

100 sq millimetres	=	1 sq centimetre
100 sq centimetres	=	1 sq decimetre
100 sq decimetres	=	1 sq metre
1000 sq metres	=	1 hectare

Capacity

1 millilitre (ml)	=	1 cubic centimetre (cm3)
1 centilitre (cl)	=	10 millilitres (ml)
1 decilitre (dl)	=	10 centilitres (cl)
1 litre (l)	=	10 decilitres (dl)

Weight

1 centigram (cg)	=	10 milligrams (mg)
1 decigram (dg)	=	10 centigrams (mcg)
1 gram (g)	=	10 decigrams (dg)
1 decagram (dag)	=	10 grams (g)
1 hectogram (hg)	=	10 decagrams (dag)

Conversion equivalents (imperial/metric)

Length

1 inch	=	25.4 mm
1 foot	=	304.8 mm
1 yard	=	914.4 mm
1 yard	=	0.9144 m
1 mile	=	1609.34 m

Area

1 sq inch	=	645.16 sq mm
1 sq ft	=	0.092903 sq m
1 sq yard	=	0.8361 sq m
1 acre	=	4840 sq yards
1 acre	=	2.471 hectares

Liquid

1 lb water	=	0.454 litres
1 pint	=	0.568 litres
1 gallon	=	4.546 litres

Horse-power

1 hp	=	746 watts
1 hp	=	0.746 kW
1 hp	=	33,000 ft.lb/min

Weight

1 lb	=	0.4536 kg
1 cwt	=	50.8 kg
1 ton	=	1016.1 kg

Conversion equivalents (metric/imperial)

Length

1 mm	=	0.03937 inches
1 centimetre	=	0.3937 inches
1 metre	=	1.094 yards
1 metre	=	3.282 ft
1 kilometre	=	0.621373 miles

Area

1 sq millimetre	=	0.00155 sq in
1 sq metre	=	10.764 sq ft
1 sq metre	=	1.196 sq yards
1 acre	=	4046.86 sq m
1 hectare	=	0.404686 acres

Liquid

1 litre	=	2.202 lbs
1 litre	=	1.76 pints
1 litre	=	0.22 gallons

Horse-power

1 watt		=	0.00134 hp
1 kw		=	134 hp
1 hp		=	0759 kg m/s

Weight

1 kg		=	2.205 lbs
1 kg		=	0.01968 cwt
1 kg		=	0.000984 ton

Temperature equivalents

In order to convert Fahrenheit to Celsius deduct 32 and multiply by 5/9.
To convert Celsius to Fahrenheit multiply by 9/5 and add 32.

Fahrenheit	Celsius
230	110.0
220	104.4
210	98.9
200	93.3
190	87.8
180	82.2
170	76.7
160	71.1
150	65.6
140	60.0
130	54.4
120	48.9
110	43.3
100	37.8
90	32.2
80	26.7
70	21.1
60	15.6
50	10.0
40	4.4
30	-1.1
20	-6.7
10	-12.2
0	-17.8

Areas and volumes

Figure	Area	Perimeter
Rectangle	Length × breadth	Sum of sides
Triangle	Base × half of perpendicular height	Sum of sides
Quadrilateral	Sum of areas of contained triangles	Sum of sides
Trapezoidal	Sum of areas of contained triangles	Sum of sides
Trapezium	Half of sum of parallel sides × perpendicular height	Sum of sides
Parallelogram	Base × perpendicular height	Sum of sides
Regular polygon	Half sum of sides × half internal diameter	Sum of sides
Circle	pi × radius²	pi × diameter or pi × 2 × radius

Figure	Surface area	Volume
Cylinder	pi × 2 × radius² × length (curved surface only)	pi × radius² × length
Sphere	pi × diameter²	Diameter3 × 0.5236

Weights of materials	kg/m3	kg/m2	kg/m
Aggregate, coarse	1,500		
Ashes	800		
Ballast	600		
Blockboard, standard	940-1000		
Blockboard, tempered	940-1060		
Blocks, natural aggregate			
75mm		160.00	
100mm		215.00	
140mm		300.00	

Weights of materials	kg/m3	kg/m2	kg/m
Blocks, lighweight aggregate			
75mm		60.00	
100mm		80.00	
140mm		112.00	
Bricks, Fletton		1820.00	
Bricks, engineering		2250.00	
Bricks, concrete		1850.00	
Brickwork, 112.5mm		220.00	
Brickwork, 215mm		465.00	
Brickwork, 327.5mm		710.00	
Cement	1,440		
Chalk	2,240		
Chipboard, standard	650-750		
Chipboard, flooring	680-800		
Clay	1,800		
Concrete	2,450		
Copper pipes, table X			
6mm			0.091
8mm			0.125
10mm			0.158
12mm			0.191
15mm			0.280
18mm			0.385
22mm			0.531
28mm			0.681
35mm			1.133
42mm			1.368
54mm			1.769
Copper pipes, table Y			
6mm			0.117
8mm			0.162
10mm			0.206
12mm			0.251
15mm			0.392
18mm			0.476
22mm			0.697
28mm			0.899
35mm			1.409
42mm			1.700
54mm			2.905

Weights of materials	kg/m3	kg/m2	kg/m
Copper pipes, table Z			
6mm			0.077
8mm			0.105
10mm			0.133
12mm			0.161
15mm			0.203
18mm			0.292
22mm			0.359
28mm			0.459
35mm			0.670
42mm			0.922
54mm			1.334
Flint	2,550		
Gravel	1,750		
Hardcore	1,900		
Hoggin	1,750		
Glass, clear sheet			
3mm		7.50	
4mm		10.00	
5mm		12.50	
6mm		15.00	
10mm		25.00	
12mm		30.00	
15mm		37.50	
19mm		47.50	
25mm		63.50	
Glass, float			
3mm		7.50	
4mm		10.00	
5mm		12.50	
6mm		15.00	
Glass, patterned			
3mm		6.00	
4mm		7.50	
5mm		9.50	
6mm		11.50	
10mm		21.50	
Laminboard	500-700		
Lime, ground	750		

Weights of materials	kg/m3	kg/m2	kg/m
Mild steel flat bars			
25 × 9.53mm			1.910
38 × 9.53mm			2.840
50 × 12.70			5.060
50 × 19.00			7.590
Mild steel round bars			
6mm			0.222
8mm			0.395
10mm			0.616
12mm			0.888
16mm			1.579
20mm			2.466
25mm			3.854
32mm			6.313
40mm			9.864
50mm			15.413
Mild steel square bars			
6mm			0.283
8mm			0.503
10mm			0.784
12mm			1.131
16mm			2.010
20mm			3.139
25mm			4.905
32mm			8.035
40mm			12.554
50mm			19.617
Plaster			
Carlite browning.11mm thick		7.80	
Carlite tough coat, 11mm thick		7.80	
Carlite bonding, 8mm thick		7.10	
Carlite bonding, 11mm thick		9.80	
Thistle hardwall, 11mm thick		8.80	
Thistle dri-coat, 11mm thick		8.30	
Thistle renovating, 11mm thick		8.80	
Sand	1,600		
Screed, 12.5mm thick		29.00	
Stone, Bath	2,200		
Stone, crushed	1,350		
Stone, Darley Dale	2,400		

Weights of materials	kg/m3	kg/m2	kg/m
Stone, natural	2,400		
Stone, Portland	2,200		
Stone, reconstructed	2,250		
Stone, York	2,400		
Terrazzo, 25mm thick		45.50	
Timber			
Ash	800		
Baltic Spruce	480		
Beech	815		
Birch	720		
Box×	960		
Cedar	480		
Ebony	1,215		
Elm	625		
Greenheart	960		
Jarrah	815		
Maple	750		
Pine, Pitchpine	800		
Pine, Red Deal	575		
Pine, Yellow Deal	530		
Sycamore	530		
Teak, African	960		
Teak, Indian	655		
Walnut	495		
Top soil	1,000		
Water	950		
Woodblock flooring			
softwood		12.70	
hardwood		17.60	
Zinc sheeting		4.6	

EXCAVATION AND FILLING

Shrinkage of deposited material

Clay	-10%
Gravel	-7.50%
Sandy soil	-12.50%

Bulking excavated material

Clay	40%
Gravel	25%
Sand	20%

Typical fuel comsumption for plant	Engine size kW	Litres per hour
Compressors up to	20	4.00
	30	6.50
	40	8.20
	50	9.00
	75	16.00
	100	20.00
	125	25.00
	150	30.00
Concrete mixers up to	5	1.00
	10	2.40
	15	3.80
	20	5.00
Dumpers	5	1.30
	7	2.00
	10	3.00
	15	4.00
	20	4.90
	30	7.00
	50	12.00
Excavators	10	2.50
	20	4.50
	40	9.00
	60	13.00
	80	17.00
Pumps	5	1.10
	10	2.10
	15	3.20
	20	4.20
	25	5.50

CONCRETE WORK

Concrete mixe	Mix	Cement t	Sand m3	Aggregate m3	Water litres
	1:1:2	0.50	0.45	0.70	208.00
	1:1.5:3	0.37	0.50	0.80	185.00
	1:2:4	0.30	0.54	0.85	175.00
	1:2.5:5	0.25	0.55	0.85	166.00
	1:3:6	0.22	0.55	0.85	160.00

BRICKWORK AND BLOCKWORK

Bricks per m2 (brick size 215 × 103.5 × 65mm)

Half brick wall
    stretcher bond   59
    English bond   89
    English garden wall bond   74
    Flemish bond   79

One brick wall
    English bond   118
    Flemish bond   118

One and a half brick wall
    English bond   178
    Flemish bond   178

Two brick wall
    English bond   238
    Flemish bond   238

Metric modular bricks

    200 × 100 × 75mm
        90mm thick   133
        190mm thick   200

200 × 100 × 100mm	
90mm thick	50
190mm thick	100
290mm thick	150

300 × 100 × 75mm	
90mm thick	44

300 × 100 × 100mm	
90mm thick	50

Blocks per m2 (block size 414 × 215mm)

60mm thick	9.9
75mm thick	9.9
100mm thick	9.9
140mm thick	9.9
190mm thick	9.9
215mm thick	9.9

Mortar per m2	**Wirecut m3**	**1 Frog m3**	**2 Frogs m3**
Brick size 215 × 103.5 × 65mm			
Half brick wall	0.017	0.024	0.031
One brick wall	0.045	0.059	0.073
One and a half brick wall	0.072	0.093	0.114
Two brick wall	0.101	0.128	0.155

Brick size 200 × 100 × 75mm	**Solid m3**		**Perforated m3**
90mm thick	0.016		0.019
190mm thick	0.042		0.048
290mm thick	0.068		0.078

Brick size 200 × 100 × 100mm			
90mm thick	0.013		0.016
190mm thick	0.036		0.041
290mm thick	0.059		0.067

	Solid m3	Perforated m3
Brick size 200 × 100 × 100mm		
90mm thick	0.015	0.018
Block size 440 × 215mm		
60mm thick	0.004	
75mm thick	0.005	
100mm thick	0.006	
140mm thick	0.007	
190mm thick	0.008	
215mm thick	0.009	

MASONRY

Mortar per m2 of random rubble wall	m3
300mm thick wall	0.120
450mm thick wall	0.160
550mm thick wall	0.120

CARPENTRY AND JOINERY

Length of boarding required	m/m2
Board width, 75mm	13.33
Board width, 100mm	10.00
Board width, 125mm	8.00
Board width, 150mm	6.67
Board width, 175mm	5.71
Board width, 200mm	5.00

ROOFING	Lap mm	Gauge mm	Nr/m2	Battens m/m2
Clay/concrete tiles				
267 × 165mm	65	100	60.00	10.00
	65	98	64.00	10.50
	65	90	68.00	11.30

	Lap mm	Gauge mm	Nr/m2	Battens m/m2
387 × 230mm	75	300	16.00	3.20
	100	280	17.40	3.50
420 × 330mm	75	340	10.00	2.90
	100	320	10.74	3.10

Fibre slates

	Lap mm	Gauge mm	Nr/m2	Battens m/m2
500 × 250mm	90	205	19.50	4.85
	80	210	19.10	4.76
	70	215	18.60	4.65
600 × 300mm	105	250	13.60	4.04
	100	250	13.40	4.00
	90	255	13.10	3.92
	80	260	12.90	3.85
	70	263	12.70	3.77
400 × 200mm	70	165	30.00	6.06
	75	162	30.90	6.17
	90	155	32.30	6.45
500 × 250mm	70	215	18.60	4.65
	75	212	18.90	4.72
	90	205	19.50	4.88
	100	200	20.00	5.00
	110	195	20.50	5.13
600 × 300mm	100	250	13.4	4.00
	110	245	13.60	4.08

Natural slates

	Lap mm	Gauge mm	Nr/m2	Battens m/m2
405 × 205mm	75	165	29.59	8.70
405 × 255mm	75	165	23.75	6.06
405 × 305mm	75	165	19.00	5.00
460 × 230mm	75	195	23.00	6.00
460 × 255mm	75	195	20.37	5.20
460 × 305mm	75	195	17.00	5.00
510 × 255mm	75	220	18.02	4.60
510 × 305mm	75	220	15.00	4.00

	Lap mm	Gauge mm	Nr/m2	Battens m/m2
560 × 280mm	75	240	14.81	4.12
560 × 280mm	75	240	14.00	4.00
610 × 305mm	75	265	12.27	3.74

Reconstructed stone slates

	Lap mm	Gauge mm	Nr/m2	Battens m/m2
380 × 250mm	75	150	16.00	3.20

PLASTERING AND TILING

Plaster coverage	m2 per 1000kg
Carlite browning, 11mm thick	135-155
Carlite tough coat, 11mm thick	135-150
Carlite bonding, 11mm thick	100-115
Thistle hardwall, 11mm thick	115-130
Thistle dri-coat, 11mm thick	135-135
Thistle renovating, 11mm thick	115-125

Tile coverage	Nr
152 × 152mm	43.27
200 × 200mm	25.00

PLUMBING AND HEATING

Roof drainage	Area m2	Pipe mm	Gutter mm
One end outlet	15	50	75
	38	68	100
	100	110	150
Centre outlet	30	50	75
	75	68	100
	200	110	150

PAINTING AND WALLPAPERING

Average coverage of paints m2 per litre	Timber	Plastered surfaces	Brickwork
Primer	10-12	9-11	5-7
Undercoat	10-12	11-14	6-8
Gloss	11-14	11-14	6-8
Emulsion	10-12	12-15	6-10

Wallpaper coverage per roll	Rolls nr	Wall height m	Room perimeter m
	4	2.50	8
	5	2.50	9
	5	2.50	10
	6	2.50	11
	6	2.50	12
	7	2.50	13
	7	2.50	14
	8	2.50	15
	8	2.50	16
	8	2.50	17
	9	2.50	18
	10	2.50	19
	10	2.50	20
	10	2.50	21
	11	2.50	22
	11	2.50	23
	12	2.50	24
	13	2.50	25
	13	2.50	26
	14	2.50	27
	5	2.80	8
	5	2.80	9
	5	2.80	10
	7	2.80	11
	7	2.80	12
	7	2.80	13
	8	2.80	14
	8	2.80	15

Rolls nr	Wall height m	Room perimeter m
9	2.8	16
10	2.8	17
10	2.8	18
11	2.8	19
11	2.8	20
12	2.8	21
13	2.8	22
13	2.8	23
14	2.8	24
14	2.8	25
15	2.8	26
15	2.8	27

EXTERNAL WORKS

Blocks/slabs per m2

	nr/m2
200 × 100mm	50.00
450 × 450mm	4.93
600 × 450mm	3.70
600 × 600mm	2.79
600 × 750mm	2.22
600 × 900mm	1.85

Drainage trench widths

	Under 1.5m deep mm	Over 1.5m deep mm
Pipe diameter 100mm	450	600
Pipe diameter 150mm	500	650
Pipe diameter 225mm	600	750
Pipe diameter 300mm	650	800

Volumes of filling for pipe beds (m3 per m)	50mm thick	100mm thick	150mm thick
Pipe diameter 100mm	0.023	0.045	0.068
Pipe diameter 150mm	0.026	0.053	0.079
Pipe diameter 225mm	0.030	0.060	0.090
Pipe diameter 300mm	0.038	0.075	0.113

Volumes of filling for pipe bed and haunching (m3 per m)	m3
Pipe diameter 100mm	0.117
Pipe diameter 150mm	0.152
Pipe diameter 225mm	0.195
Pipe diameter 300mm	0.279

Volumes of filling for pipe bed and surround (m3 per m)	
Pipe diameter 100mm	0.185
Pipe diameter 150mm	0.231
Pipe diameter 225mm	0.285
Pipe diameter 300mm	0.391

Part Six

BUSINESS MATTERS

Starting a business

Running a business

Taxation

Starting a business

Most small businesses come into being for one of two reasons – ambition or desperation! A person with genuine ambition for commercial success will never be completely satisfied until he has become self-employed and started his own business. But many successful businesses have been started because the proprietor was forced into this course of action because of redundancy.

Before giving up his job, the would-be businessman should consider carefully whether he has the required skills and the temperament to survive in the highly competitive self-employed market. Before commencing in business it is essential to assess the commercial viability of the intended business because it is pointless to finance a business that is not going to be commercially viable.

In the early stages it is important to make decisions such as: What exactly is the product being sold? What is the market view of that product? What steps are required before the developed product is first sold and where are those sales coming from?

As much information as possible should be obtained on how to run a business before taking the plunge. Sales targets should be set and it should be clearly established how those important first sales are obtained. Above all, do not underestimate the amount of time required to establish and finance a new business venture.

Whatever the size of the business it is important that you put in writing exactly what you are trying to do. This means preparing a business plan that will not only assist in establishing your business aims but is essential if you need to raise finance. The contents of a typical business plan are set out later. It is important to realise that you are not on your own and there are many contacts and advising agencies that can be of assistance.

Potential customers and trade contacts

Many persons intending to start a business in the construction industry will have already had experience as employees. Use all contacts to check the market, establish the sort of work that is available and the current charge-out rates.

In the domestic market, check on the competition for prices and services provided. Study advertisements for your kind of work and try to get firm promises of work before the start-up date.

Testing the market

Talk to as many traders as possible operating in the same field. Identify if the market is in the industrial, commercial, local government or in the domestic field. Talk to prospective customers and clients and consider how you can improve on what is being offered in terms of price, quality, speed, convenience, reliability and back-up service.

Business links

There is no shortage of information about the many aspects of starting and running your own business. Finance, marketing, legal requirements, developing your business idea and taxation matters are all the subject of a mountain of books, pamphlets, guides and courses so it should not be necessary to pay out a lot of money for this information. Indeed, the likelihood is that the aspiring businessman will be overwhelmed with information and will need professional guidance to reduce the risk of wasting time on studying unnecessary subjects.

Business Links are now well established and provide a good place to start for both information and advice. These organisations provide a 'one-stop-shop' for advice and assistance to owner-managed businesses. They will often replace the need to contact Training and Enterprise Councils (TECs) and many of the other official organisations listed below.
Point of contact: telephone directory for address.

Training and Enterprise Councils (TECs)

TECs are comprised of a board of directors drawn from the top men in local industry, commerce, education, trade unions etc., who, together with their staff and experienced business counsellors, assist both new and established concerns in all aspects of running a business. This takes the form of across-the-table advice and also hands-on assistance in management, marketing and finance if required. There are also training courses and seminars available in most areas together with the possibility of grants in some areas.
Point of contact: local Jobcentre or Citizens' Advice Bureau.

Banks

Approach banks for information about the business accounts and financial services that are available. Your local Business Link can advise on how best to find a suitable bank manager and inform you as to what the bank will require.

Shop around several banks and branches if you are not satisfied at first because managers vary widely in their views on what is a viable business proposition. Remember, most banks have useful free information packs to help business start-up.

Point of contact: local bank manager.

HM Inspector of Taxes

Make a preliminary visit to the local tax office enquiry counter for their publications on income tax and national insurance contributions.

SA/Bk 3	Self assessment. A guide to keeping records for the self employed
IR 15(CIS)	Construction Industry Tax Deduction Scheme˙
CWL	Starting your own business,
IR 40(CIS)	Conditions for Getting a Sub-Contractor's Tax Certificate
NE1	PAYE for Employers (if you employ someone)
NE3	PAYE for new and small Employers
IR 56/N139	Employed or Self-Employed. A guide for tax and National Insurance
CA02	National Insurance contributions for self employed people with small earnings.

Remember, the onus is on the taxpayer, within three months, to notify the Inland Revenue that he is in business and failure to do so may result in the imposition of £100 penalty. Either send a letter or use the form provided at the back of the *'Starting your own business booklet'* to the Inland Revenue National Insurance Contributions Office and they will inform your local tax office of the change in your employment status.

Point of contact: telephone directory for address.

Inland Revenue National Insurance Contributions Office

Self Employment Services
Customer Accounts Section
Longbenton
Newcastle NE 98 1ZZ

Telephone the Call Centre on 0845 9154655 and ask for the following publications:

CWL2	Class 2 and Class 4 Contributions for the Self Employed
CA02	People with Small Earnings from Self-Employment
CA04	Direct Debit - The Easy Way to Pay. Class 2 and Class 3
CA07	Unpaid and Late Paid Contributions
	and for Employers
CWG1	Employer's Quick Guide to PAYE and NIC Contributions
CA30	Employer's Manual to Statutory Sick Pay

VAT

The VAT office also offer a number of useful publications, including;

700	The VAT Guide
700/1	Should I be Registered for VAT?
731	Cash Accounting
732	Annual Accounting
742	Land and Property

Information about the Cash Accounting Scheme and the introduction of annual VAT returns are dealt with later.
Point of contact: telephone directory for address.

Local authorities

Authorities vary in provisions made for small businesses but all have been asked to simplify and cut delays in planning applications. In Assisted Areas, rent-free periods and reductions in rates may be available on certain industrial and commercial properties. As a preliminary to either purchasing or renting business premises, the following booklets will be helpful:

Step by Step Guide to Planning Permission for Small Businesses, and
Business Leases and Security of Tenure

Both are issued by the Department of Employment and are available at
council offices, Citizens' Advice Bureaux and TEC offices. Some authorities
run training schemes in conjunction with local industry and educational
establishments.

Point of contact: usually the Planning Department - ask for the Industrial
Development or Economic Development Officer.

Department of Trade and Industry

The services formally provided by the Department are now increasingly being
provided by Business Link . The Department can still, however, provide
useful information on available grants for start-ups.
Point of contact: telephone 0207-215 5000 and ask for the address and
telephone number of the nearest DTI office and copies of their explanatory
booklets.

Department of Transport and the Regions

Regulations are now in force relating to all forms of waste other than normal
household rubbish. Any business that produces, stores, treats, processes,
transports, recycles or disposes of such waste has a 'duty of care' to ensure it
is properly discarded and dealt with.
   Practical guidance on how to comply with the law (it is a criminal offence
punishable by a fine not to) is contained in a booklet *Waste Management: The
Duty of Care: A Code of Practice,* obtainable from HMSO Publication Centre,
PO Box 276, London SW8 5DT. Telephone 0207-873 9090.

Accountant

The services of an accountant are to be strongly recommended from the
beginning because the legal and taxation requirements start immediately and
must be properly complied with if trouble is to be avoided later. A qualified
accountant must be used if a limited company is being formed but an
accountant will give advice on a whole range of business issues including
book-keeping, tax planning and compliance to finance raising and will help in

preparing annual accounts.

It is worth spending some time finding an accountant who has other clients in the same line of business and is able to give sound advice particularly on taxation and business finance and is not so overworked that damaging delays in producing accounts are likely to arise. Ask other traders whether they can recommend their own accountant. Visit more than one firm of accountants, ask about the fees they charge and how much the production of annual accounts and agreement with the Inland Revenue are likely to cost. A good accountant is worth every penny of his fees and will save you both money and worry.

Solicitor

Many businesses operate without the services of a solicitor but there are a number of occasions when legal advice should be sought. In particular, no-one should sign a lease of premises without taking legal advice because a business can encounter financial difficulty through unnoticed liabilities in its lease. Either an accountant or solicitor will help with drawing up a partnership agreement which all partnerships should have. A solicitor will also help to explain complex contractual terms and prepare draft contracts if the type of business being entered into requires them.

Insurance broker

Policies are available to cover many aspects of business including:

- employer's liability - compulsory if the business has employees
- public liability - essential in the construction industry
- motor vehicles
- theft of stock, plant and money
- fire and storm damage
- personal accident and loss of profits
- keyman cover.

Brokers are independent advisers who will obtain competitive quotations on your behalf. See more than one broker before making a decision - their advice is normally given free and without obligation.

Point of contact: telephone directory or write for a list of local members to:

The British Insurance Brokers' Association
   Consumer Relations Department
   BIBA House
   14 Bevis Marks
   London
   EC3A 7NT (telephone: 0207-623 9043)
                    or contact
   The Association of British Insurers
   51 Gresham Street
   London
   EC2V 7HQ (telephone: 0207-600 3333)

who will supply free a package of very useful advice files specially designed
for the small business.

The Health and Safety Executive

The Executive operates the legislation covering everyone engaged in work
activities and has issued a very useful set of '*Construction Health Hazard
Information Sheets*' covering such topics as handling cement, lead and
solvents, safety in the use of ladders, scaffolding, hoists, cranes, flammable
liquids, asbestos, roofs and compressed gases etc. A pack of these may be
obtained free from your local HSE office or The Health & Safety Executive
Central Office, Sheffield (telephone: (01142-892345) or HSE Publications
(telephone: 01787-881165).

Business plan

As stated before, once the relevant information has been obtained it should be
consolidated into a formal business plan. The complexity of the plan will
depend in the main on the size and nature of the business concerned.
Consideration should be given to the following points.

Objectives

It is important to establish what you are trying to achieve both for you and the
business. A provider of finance may be particularly influenced by your ability
to achieve short- and medium-term goals and may have confidence in
continuing to provide finance for the business. From an individual point of

view, it is important to establish goals because there is little point in having a business that only serves to achieve the expectations of others whilst not rewarding the would-be businessman.

History

If you already own an existing business then commentary on its existing background structure and history to date can be of assistance. There is no substitute for experience and any existing contacts you have in the construction industry will be of assistance to you. The following points should also be considered for inclusion:

- a brief history of the business identifying useful contacts made
- the development of the business, highlighting significant successes and
  their relevance to the future
- principal reasons for taking the decision to pursue this new venture
- details of present financing of the business.

Products or services

It is important to establish precisely what it is you are going to sell. Does the product or service have any unique qualities which gives it your advantages over competitors? For example, do you have an ability to react more quickly than your competitors and are you perceived to deliver a higher quality product or service?
A typical business plan would include:

- description of the main products and services
- statement of disadvantages and advising how they will be overcome
- details of new products and estimated dates of introduction
- profitability of each product
- details of research and development projects
- after-sales support.

Markets and marketing strategy

This section of the business plan should show that thought has been given to the potential of the product. In this regard it can often be useful to identify major
competitors and make an overall assessment of their strengths and weaknesses, including the following:

- an overall assessment of the market, setting out its size and growth potential
- a statement showing your position within the market
- an identification of main customers and how they compare
- details of typical orders and buying habits
- pricing strategy
- anticipated effect on demand of pricing
- expectation of price movement
- details of promotions and advertising campaigns.

It is important to identify your customers and why they might buy from you. Those entering the domestic side of the business will need to think about the best way to reach potential customers. Are local word-of-mouth recommendations enough to provide reasonable work continuity. If not, what is the most effective method of advertising to reach your customer base?
    Remember, advertising is costly. It is a waste of funds to place an advertisement in a paper circulating in areas A, B, C & D if the business only covers area A.

Research and development

If you are developing a product or a particular service, then an assessment should be made on what stage it is at and what further finance is required to complete it. It may also be useful to make an assessment on the vulnerability of the product or service to innovations being initiated by others.

Basis of operation

Detail what facilities you will require in order to carry on your trade in the form of property, working and storage areas, office space, etc. An assessment should also he made on the assistance you will require from others. Your business plan might include:

- a layman's guide to the process or work
- details of facilities, buildings and plant
- key factors affecting production, such as yields and wastage
- raw material demand and usage.

Management

This section is one of the most important because it demonstrates the capability of the would-be businessman. The skills you need will cover production, marketing, finance and administration. In the early stages you may be able to do this yourself but as the business grows it may be required to develop a team to handle these matters. The following points should be considered for inclusion in the plan:

- set out age, experience and achievements
- state additional management requirements in the future and how they are to be met
- identify current weaknesses and how they will be overcome
- state remuneration packages and profit expectations
- give detailed CVs in appendices.

Advertising and retraining may be required in order to identify and provide suitable personnel where expertise and experience are lacking.

Financial information

It is important to detail, if any, the present financial position of your business and the budgeted profit and loss accounts, cash flows and balance sheets. These integrated forecasts should be prepared for the next twelve months at monthly intervals and annually for the following two years.

   If the forecasts are to be reasonably accurate then the businessman must make some early decisions about:

- the premises where the business will be based, the initial repairs and alterations that might he required and an assessment of the total cost

- which plant, equipment and transport are needed, whether they are to be leased or purchased and what the cost will be?
- how much stock of materials, if any, should be carried? - the bare minimum only should be acquired, so reliable suppliers should be found
- what will be the weekly bills for overheads, wages and the proprietor's living costs?
- what type of work is going to be undertaken, and how much profit can realistically be obtained?
- how often are invoices to be presented?

Your business plan should include the following information:

- explanation of how sales forecasts are prepared
- levels of production
- details of major variable overheads and estimates
- assumptions in cash flow forecasting, inflation and taxation.

Finance required and its application

The financial details given above should produce an accurate assessment of the funds required to finance the business. It is important to distinguish between those items that require permanent finance and those that will eventually be converted to cash because it is not usually advisable to finance long-term assets with personal equity.

Working capital such as stock and debtors can usually be obtained by an overdraft arrangement but your accountant or bank will advise you on this.

Executive summary

Although it is prepared last, this summary will be the first part of your business plan. Remember that business plans are prepared for busy people and their decision on finance may be based solely on this section. It should cover two or three pages and deal with the most important aspects and opportunities in your plan. Here are some of the main headings:

- key strategies
- finance required and how it is to be used
- management experience

- anticipated returns and profits
- markets.

The appendices should include:

- CVs of key personnel
- organisation charts
- market studies
- product advertising literature
- professional references
- financial forecasts
- glossary of terms.

If you feel that any additional information should be provided in support of your proposal, then this is usually best included in the appendices.

Follow up

Please remember that once your plan is prepared, it is important to re-examine it regularly and update the forecasts and financial information. This is a working document and can be an important tool in running the business.

Sources of finance

Personal funds

Finance, like charity, often begins at home and a would-be businessman should make a realistic assessment of his net worth, including the value of his house after deducting the mortgage(s) outstanding on it, savings, any car or van owned and any sums which the family are prepared to contribute but deducting any private borrowings which will come due for payment. The whole of these funds may not be available (for instance, money which has been loaned to a friend or relative who is known to be unable to repay at the present time).

It may not be desirable that all capital should be put at risk on a business venture so the following should be established:

- how much cash you propose to invest in the business
- whether the family home will be made available for any business borrowing
- state total finance required
- how finance is anticipated being raised
- interest and security to be provided
- expected return on investment.

Whilst it may be wise not to pledge too much of the family assets, it has to be remembered that the bank will be looking closely at the degree to which the proprietor has committed himself to the venture and will not be impressed by an application for a loan where the applicant is prepared to risk only a small fraction of his own resources.

Having decided how much of his own funds to contribute, the businessman can now see the level of shortfall and consider how best to fill it. Consideration should be given to partners where the shortfall is large and particularly when there is a need for heavy investment in fixed assets, such as premises and capital equipment. It may be worthwhile starting a limited company with others also subscribing capital and to allow the banks to take security against the book debts.

Banks

The first outside source of money to which most businessmen turn is the bank and here are a few guidelines on approaching a bank manager:

- present your business plan to him; remember to use conservative estimates which tend to understate rather than overstate the forecast sales and profits
- know the figures in detail and do not leave it to your accountant to explain them for you. The bank manager is interested in the businessman not his advisers and will be impressed if the businessman demonstrates a grasp of the financing of his business
- understand the difference between short- and long-term borrowing

- ask about the Government Loan Guarantee Scheme if there is a shortage of security for loans. The bank may be able to assist, or depending on certain conditions being met, the Government may guarantee a certain percentage of the bank loan.

Remember the bank will want their money back, so bank borrowings are usually required to be secured by charges on business assets. In start-up situations, personal guarantees from the proprietors are normally required. Ensure that if these are given they are regularly reviewed to see if they are still required.

Enterprise Investment Scheme - business angels

If an outside investor is sought in a business he will probably wish to invest within the terms of the Enterprise Investment Scheme which enables him to gain income tax relief at 20% on the amount of his investment. Additionally, any investment can be used to defer capital gains tax. The rules are complex and professional advice should always be sought.

Hire purchase/leasing

It is not always necessary to purchase assets outright that are required for the business and leasing and hire purchase can often form an integral part of a business's medium-term finance strategy.

Venture capital

In addition, there are a number of other financial institutions in the venture capital market that can help well-established businesses, usually limited companies, who wish to expand. They may also assist well-conceived start-ups. They will provide a flexible package of equity and loan capital but only for large amounts, usually sums in excess of £150,000 and often £250,000.

Usually the deal involves the financial institution having a minority interest in the voting share capital and a seat on the board of the company. Arrangements for the eventual purchase of the shares held by the finance company by the private shareholders are also normally incorporated in the scheme.

The Royal Jubilee and Princes Trust

These trusts through the Youth Business Initiative provide bursaries of not more than £1,000 per individual to selected applicants who are unemployed and age 25 or over. Grants may be used for tools and equipment, transport, fees, insurance, instruction and training but not for working capital, rent and rates, new materials or stock. They operate through a local representative whose name and address may be ascertained by contacting the Prince's Youth Business.
Point of contact: telephone 0207-321 6500.

The Business Start-up Scheme

This is an allowance of £50 per week, in addition to any income made from your business, paid for twenty weeks. To qualify you must be at least 18 and under 65, work at least 36 hours per week in the business and have been unemployed for at least six months or fall into one of the other categories: disabled, ex-HMS or redundant.

The first step is to get the booklet on the subject from your local Jobcentre or TEC that includes details on how and where to apply. Once in receipt of the enterprise allowance, you will also have the benefit of advice and assistance from an experienced businessman from your TEC. All the initial counselling services and training courses are free.

Running a business

Many businesses are run without adequate information being available to check trend in their vital areas, e.g. marketing, money and managerial efficiency. It is essential to look critically at all aspects of the business in order to maximise profits and reduce inefficiency. Regular meaningful information is required on which management can concentrate. This will vary according to the proprietor's business but will often concentrate on debtors, creditors, cash, sales and orders.

Proprietors often have the feeling that the business should be 'doing better' but are unable to identify what is going wrong. Sometimes there is the worrying phenomenon of a steadily increasing work programme coupled with a persistently reducing bank balance or rising overdraft. Some useful ways of checking the position and of identifying problem areas are given below.

Marketing

Throughout his business life the entrepreneur should continuously study the methods and approach of his competitors. A shortcoming frequently found in ailing concerns is that the proprietor thinks he knows what his customers want better than they do.

The term 'market research' sounds both difficult and expensive but a very simple form of it can be done quite effectively by the businessman and his sales staff. Existing and prospective customers should be approached and asked what they want in terms of price, quality, design, payment terms, follow-up service, guarantees and services.

The initial approach might be by a leaflet or letter followed by a personal call. As an on-going part of management, all staff with customer contact should be encouraged to enquire about and record customer preferences, complaints, etc. and feed it back to management.

Other sources of information can be trade and business journals, trade exhibitions, suppliers and representatives from which information about trends, new techniques and products can be obtained and studied. Valuable information can also be gained from studying competitors and the following questions should be asked:

- what do they sell and at what prices?

-    what inducements do they offer to their customers, e.g. credit
     facilities, guarantees, free offers and discounts?
-    how do they reach their customers - local/national advertising,
     mail shots, salesmen, local radio and TV?
-    what are the strongest aspects of their appeal to customers and
     have they any weaknesses?

The businessman should apply all the information gathered from customers
and competitors to his own services with a view to making sure he is offering
the right product at the right price in the most attractive way and in the most
receptive market.

In a small business where the proprietor is also his own salesman he must
give careful thought on how he can best present his product and himself. For
instance, if he is working solely within the construction industry his main
problems are likely to centre on getting a C1S6 Certificate and using trade
contacts to get sub-contract work.

However, for those who serve the general public, presentation can be a vital
element in getting work. The customer is looking for efficiency, reliability
and honesty in a trader and quality, price and style in the product. To bring
out these facets in discussion with a potential customer is a skilled task. A
short course on marketing techniques could pay handsome dividends. The
Business Link will give the names and addresses of such courses locally.

Financial control

Unfortunately, some unsuccessful firms do not seek financial advice until too
late when the downward trend cannot be halted. Earlier attention to the
problems may have saved some of them so it is important to recognise the
tell-tale signs. There are some tests and checks that can be done quite easily.

Cash flow

Cash flow is the lifeblood of the business and more businesses fail through
lack of cash than for any other reason. Cash is generated through the
conversion of work into debtors and then into payment and also through the
deferral of the payment of supplies for as long a period that can be negotiated.
The objective must be to keep stock, work in progress, debts to a minimum
and creditors to a maximum.

Debtor days

This is calculated by dividing your trade debtors by annual sales and multiplying by 365. This shows the number of days' credit being afforded to your customers and should be compared both with your normal trade terms and the previous month's figures. Normal procedures should involve the preparation of a monthly-aged list of debtors showing the name of the customer, the value and to which month it relates.

The oldest and largest debtors can be seen at a glance for immediate consideration of what further recovery action is needed. The list may also show over-reliance on one or two large customers or the need to stop supplying a particularly bad payer until his arrears have been reduced to an acceptable level. Consideration should be given to making up bills to a date before the end of the month and making sure the accounts are sent out immediately, followed by a statement four weeks later.

Consider giving discounts for prompt payment. If all else fails, and legal action for recovery is being contemplated, call at the County Court and ask for their leaflets.

Stock turn

The level of stock should be kept to a minimum and the number of days' stock can be calculated by dividing the stock by the annual purchases and multiplying by 365. A worsening trend on a month-by-month basis shows the need for action. It is important to regularly make a full inventory of all stock and dispose of old or surplus items for cash. A stock control procedure to avoid stock losses and to keep stock to a minimum should be implemented.

Profitability

Whilst cash is vital in the short-term, profitability is vital in the medium-term. The two key percentage figures are the gross profit percentage and the net profit percentage. Gross profit is calculated by deducting the cost of materials and direct labour from the sales figures whilst net profit is arrived at after deducting all overheads. Possible reasons for changes in the gross profit percentage are:

-   not taking full account of increases in materials and wages in the pricing of jobs

- too generous discount terms being offered
- poor management, over-manning, waste and pilferage of materials
- too much down-time on equipment which is in need of replacement.

If net profit is deteriorating after the deduction of an appropriate reward for your own efforts, including an amount for your own personal tax liability, you should review each item of overhead expenditure in detail asking the following questions:

- can savings be made in non-productive staff?
- is sub-contracting possible and would it be cheaper?
- have all possible energy-saving methods been fully explored?
- do the company's vehicles spend too much time in the yard and can they be shared or their number reduced?
- is the expenditure on advertising producing sales - review in association with 'marketing' above?

Over-trading

Many inexperienced businessmen imagine that profitability equals money in the bank and in some cases, particularly where the receipts are wholly in cash, this may be the case. But often, increased business means higher stock inventories, extra wages and overheads, increased capital expenditure on premises and plant, all of which require short-term finance.

Additionally, if the debtors show a marked increase as the turnover rises, the proprietor may find to his surprise that each expansion of trade reduces rather than increases his cash resources and he is continually having to rely on extensions to his existing credit.

The business, which had enough funds for start-up, finds it does not have sufficient cash to run at the higher level of operation and the bank manager may he getting anxious about the increasing overdraft. It is essential for those who run a business that operates on credit terms to be aware that profitability does not necessarily mean increased cash availability. Regular monthly management information on marketing and finance as described in this chapter will enable over-trading to be recognised and remedial action to be taken early.

If the situation is appreciated only when the bank and other creditors are pressing for money, radical solutions may be necessary, such as bringing in new finance, sale and leaseback of premises, a fundamental change in the terms of trade or even selling out to a buyer with more resources. Professional help from the firm's accountant will be needed in these circumstances.

Break-even point

The costs of a business may be divided into two types - variable and fixed. *Variable costs* are those which increase or decrease as the volume of work goes up or down and include such items as materials used, direct labour and power machine tools. *Fixed costs* are not related to turnover and are sometimes called fixed overheads. They include rent, rates, insurance, heat and light, office salaries and plant depreciation. These costs are still incurred even though few or no sales are being made.

Many small businessmen run their enterprises from home using family labour as back-up; they mainly sell their own labour and buy materials and hire plant only as required. By these means they reduce their fixed costs to a minimum and start making profits almost immediately. However, larger firms that have business premises, perhaps a small workshop, an office and vehicles, need to know how much they have to sell to cover their costs and become profitable.

In the case of a new business it is necessary to estimate this figure but where annual accounts are available a break-even chart based on them can be readily prepared. Suppose the real or estimated figures (expressed in £000s) are:

	%	£
Sales	100	400
Variable costs	66	265
Gross profit	34	135
Fixed costs	13	50
Net profit	21	85

Break-even point = $\dfrac{50}{(1 \text{ less variable costs } \%)}$ sales

$$= \quad \text{50 divided by (1 less 0.6625)}$$
$$= \quad \text{50 divided by 0.3375}$$
$$= \quad \text{£148 (thousand)}$$

In practice things are never quite as clear cut as the figures show, but nevertheless this is a very useful tool for assessing not only the break-even point but also the approximate amount of loss or profit arising at differing levels of turnover and also for considering pricing policy.

Taxation

The first decision usually required to be made from a taxation point of view is which trading entity to adopt. The options available are set out below.

Sole trader

A sole trader is a person who is in business on his own account. There is no statutory requirement to produce accounts nor is there a necessity to have them audited. A sole trader may, however, be required to register for PAYE and VAT purposes and maintain records so that Income Tax and VAT returns can be made. A sole trader is personally liable for all the liabilities of his business.

Partnership

A partnership is a collection of individuals in business on their own account and whose constitution is generally governed by the Partnership Act 1890. It is strongly recommended that a partnership agreement is also established to determine the commercial relationship between the individuals concerned.

    The requirements in relation to accounting records and returns are similar to those of a sole trader and in general a partner's liability is unlimited.

Limited company

This is the most common business entity. Companies are incorporated under the Companies Act 1985 which requires that an annual audit is carried out for all companies with a turnover in excess of £1,000,000 or a review if the turnover is less than £1,000,000 and that accounts are filed with the Companies Registrar. Generally an individual shareholder's liability is limited to the amount of the share capital he is required to subscribe.

Advantages

In view of the problems and costs of incorporating an existing business, it is important to try and select the correct trading medium at the commencement of operations. It is not true to say that every business should start life as a company.

Many businesses are carried on in a safe and efficient manner by sole traders or partnerships. Whilst recognising the possible commercial advantages of a limited company, taxation advantages exist for sole traderships and partnerships, such as income tax deferral and National Insurance saving. No decision should be taken without first seeking professional advice.

The benefit of limited liability should not be ignored although this can largely be negated by banks seeking personal guarantees. In addition, it may be easier for the companies to raise finance because the bank can take security on the debts of the company that could be sold in the future, particularly if third-party finance has been obtained in the form of equity.

Self-assessment

From the tax year 1996/97 the burden of assessing tax shifted from the Inland Revenue to the individual tax payer. The main features of this system are as follows:

- the onus is on the taxpayer to provide information and to complete returns
- tax will be payable on different dates
- the taxpayer has a choice: he can calculate his tax liability at the same time as making his return and this will need to be done by 31st January following the end of the tax year. Alternatively, he can send in his tax return before 30 September and the Inland Revenue will calculate the tax to be paid on the following 31 January
- the important aspect to the system is that if the return is late, or the tax is paid late, there will be automatic penalties and/or surcharges imposed on the taxpayer.

Tax correspondence

Businessmen do not like letters from the Inland Revenue but they should resist the temptation to tear them up or put them behind the clock and forget about them. All Tax Calculations and Statements of Account should be checked for accuracy immediately and any queries should be put to your accountant or sent to the Tax District that issued the document.

Keep copies of all correspondence with the Inland Revenue. Letters can be mislaid or fail to be delivered and it is essential to have both proof of what was sent as well as a permanent record of all correspondence.

Dates tax due

Income Tax
Payments on account (based on one half of last year's liability) are due on 31 January and 31 July. If these are insufficient there is a balancing payment due on the following 31 January – the same day as the tax return needs to be filed. For example:

for the year 2000/01   Tax due £5,000 (1999/00 was £4,000)
First payment on account of £2,000 is due on 31.01.01
Second payment on account of £2,000 is due on 31.07.01
Balancing payment of £1,000 is due on 31.01.02

Note that on 31.01.02, the first payment on account of £2,500 fell due for the tax year 2001/02.

Tax in business

Spouses in business

If spouses work in the business, perhaps answering the phone, making appointments, writing business letters, making up bills and keeping the books, they should be properly remunerated for it. Being a payment to a family member, the Inspector of Taxes will be understandably cautious in allowing remuneration in full as a business expense. The payment should be:

- actually paid to them, preferably weekly or monthly and in addition to any housekeeping monies
- recorded in the business book
- reasonable in amount in line with their duties and the time spent on them.

If the wages paid to them exceed £77.00 per week, Class 1 employer's and employee's NIC becomes due and if they exceed £4,155 p.a. (assuming they have no other income) PAYE tax will also be payable.

It should also be noted that once small businesses are well established and the spouses' earnings are approaching the above limits, consideration may be given to bringing them in as a partner. This has a number of effects:

- there is a reduced need to relate the spouse's income (which is now a share of the profits) to the work they do
- they will pay Class 2 and Class 4 NIC instead of the more costly Class I contributions and PAYE will no longer apply to their earnings but remember that, as partners, they have unlimited liability.

Premises

Many small businessmen cannot afford to rent or buy commercial premises and run their enterprises from home using part of it as an office where the books and vouchers, clients' records and trade manuals are kept and where estimates and plans are drawn up. In these circumstances, a portion of the outgoings on the property may be claimed as a business expenses. An accountant's advice should be sought to ensure that the capital gains tax exemption that applies on the sale of the main residence is not lost.

Fixed Profit Car Scheme

It may be advantageous to calculate your car expenses using a fixed rate per business mile. A condition is that your annual turnover is below the VAT threshold (currently £55,000). Ask your accountant about this. A proper record of business mileage must be kept.

Vehicles

Car expenses for sole traders and partners are usually split on a fractional mileage basis between business journeys, which are allowable, and private ones, which are not, and a record of each should he kept. If the business does work only on one or two sites for only one main contractor, the inspector may argue that the true base of operations is the work site not the residence and seek to disallow the cost of travel between home and work. It is tax-wise and sound business practice to have as many customers as possible and not work for just one client.

Business entertainment

No tax relief is due for expenditure on business entertainment and neither is the VAT recoverable on gifts to customers, whether they are from this country or overseas. However, the cost of small trade gifts not exceeding £10 per person per annum in value is still admissible provided that the gift advertises the business and does not consist of food, drink or tobacco.

Income tax (2003/04)

Personal allowances

The current personal allowance for a single person is £4,155. The personal allowance for people aged 65 to 74 and over 75 years are £6,610 and £6,720 respectively. The married couple's allowance was withdrawn on 5 April 2000, except for those over 65 on that date.

Taxation of husband and wife

A married woman is treated in much the same way as a single person with her own personal allowance and basic rate band. Husband and wife each make a separate return of their own income and the Inland Revenue deals with each one in complete privacy; letters about the husband's affairs will be addressed only to him and about the wife's only to her unless the parties indicate differently.

Rates of tax

Tax is deducted at source from most banks and building societies accounts at the rate of 20%. The rates of tax for 2000/01 are as follows:

Lower rate: 10% on taxable income up to £1,960
Basic rate: 22% on taxable income between £1,960 and £30,500
Higher rate: 40% on taxable income over £30,500

Dividends carry a 10% non-repayable tax credit. Higher rate taxpayers pay a further tax on dividends of 22.5%.

Mortgage interest relief

This is no longer available after 5 April 2000.

Business losses

These are allowed only against the income of the person who incurs the loss. For example, a loss in the husband's business cannot be set against the wife's income from employment.

Joint income

In the case of joint ownership by a husband and wife of assets that yield income, such as bank and building society accounts, shares and rented property, the Inland Revenue will treat the income as arising equally to both and each will pay tax on one half of the income. If, however, the asset is owned in unequal shares or one spouse only and the taxpayer can prove this, then the shares of income to be taxed can be adjusted accordingly if a joint declaration is made to the tax office setting out the facts.

Capital Gains Tax

Where an asset is disposed of, the first £7,900 of the gain is exempt from tax. In the case of husbands and wives, each has a £7,900 exemption so if the ownership of the assets is divided between them, it is possible to claim exemption on gains up to £15,800 jointly in the tax year. Any remaining gain is chargeable as though it were the top slice of the individual's income; therefore according to his or her circumstances it might be charged at 10%, 22% or 40%.

Self-employed NIC rates (from 6 April 2000)

Class 2 rate
Charged at £2.00 per week. If earnings are below £4,615 per annum averaged over the year, ask the DSS about 'small income exception'. Details are in leaflet CA02.

Class 4 rate
Business profits up to £4,615 per annum are charged at NIL. Annual profits
between £4,615 and £30,940 are charged at 8% of the profit. There is a charge
on profits over £30,940. Class 4 contributions are collected by the Inland
Revenue along with the income tax due.

Capital allowances (depreciation) rates

Plant and machinery:	25% (40% first-year allowance is available for certain small businesses)
Business motor cars - cost up to £12,000:	25%
- cost over £12,000:	£3,000 (maximum)
Industrial build	4%
Commercial and industrial buildings in Enterprise Zones:	100%
Computers and software equipment	100%

THE CONSTRUCTION INDUSTRY TAX DEDUCTION SCHEME

General

The new Construction Industry Tax Deduction Scheme is known as the 'CIS'
scheme and replaced the old '714' scheme. As the scheme operates whenever
a contractor makes a payment to a sub-contractor, the businessman should
visit his local income tax enquiry office and obtain copies of the Inland
Revenue booklet IR 14/15 (CIS) and leaflet IR 40 which will explain the
conditions under which the Inland Revenue will issue a registration card or
(CIS6) certificate and precisely when the scheme applies.

Everyone who carries out work in the Construction Industry Scheme must
hold a registration card (CIS4) or a tax certificate (CIS6). Certain larger
companies use a special certificate (CIS5).

If the sub-contractor has a registration card but does not hold a valid tax certificate (CIS6) issued to him by the Inland Revenue, then the contractor *must* deduct 18% tax from the whole of any payment made to him (excluding the cost of any materials) and to account to the Inland Revenue for all amounts so withheld.

To enable the subcontractor to prove to the Inspector of Taxes that he has suffered this tax deduction, the contractor must complete the three-part tax payment voucher (CIS25) showing the amount withheld. These vouchers must be carefully filed for production to the Inspector after the end of the tax year along with the tax return. Any tax deducted in this way over and above the sub-contractor's agreed liability for the year will be repaid by the Inland Revenue. If he holds a (CIS6) certificate the payment may be made in full without deducting tax.

A small business that does work only for the general public and small commercial concerns is outside the scheme and does not need a certificate to trade. If, however, it engages other contractors to do jobs for it, the business would have to register under the scheme as a contractor and deduct tax from any payment made to a sub-contractor who did not produce a valid (CIS6) certificate. If in doubt, consult your accountant or the Inland Revenue direct.

VAT

The general rule about liability to register for VAT is given in the VAT office notes. It is possible to give here only a brief outline of how the tax works. The rules that apply to the construction industry are extremely complex and all traders must study *The VAT Guide* and other publications.

Registration for VAT is required if, at the end of any month, the value of taxable supplies in the last 12 months exceeds the annual threshold or if there are reasonable grounds for believing the value of the taxable supplies in the next 30 days will exceed the annual threshold.

Taxable supplies include any zero-rated items. The annual threshold is £55,000. The amount of tax to be paid is the difference between the VAT charged out to customers *(output tax)* and that suffered on payments made to suppliers for goods and services *(input tax)* incurred in making taxable supplies. Unlike income tax there is no distinction in VAT for capital items so that the tax charged on the purchase of, for example, machinery, trucks and office furniture, will normally be reclaimable as *input tax*.

VAT is payable in respect of three monthly periods known as 'tax periods'. You can apply to have the group of tax periods that fits in best with your financial year. The tax must be paid within one month of the end of each tax period. Traders who receive regular repayments of VAT can apply to have them monthly rather than quarterly. Not all types of goods and services are taxed at 17.5% (i.e. the standard rate). Some are exempt and others are zero-rated.

Zero-rated

This means that no VAT is chargeable on the goods or services, but a registered trader can reclaim any *input* tax suffered on his purchases. For instance, a builder pays VAT on the materials he buys to provide supplies of constructing but if he is constructing a new dwelling house, this is zero rated. The builder may reclaim this VAT or set it off against any VAT due on standard rated work.

Exempt

Supplies that are exempt are less favourably treated than those that are zero rated. Again no VAT is chargeable on the goods or services but the trader cannot reclaim any *input* tax suffered on his purchases.

Standard-rated

All work which is not specifically stated to be zero rated or exempt is standard-rated, i.e. VAT is chargeable at the current rate of 17.5% and the trader may deduct any *input* tax suffered when he is making his return to the Customs and Excise. If for any reason a trader makes a supply and fails to charge VAT when he should have done so (e.g. mistakenly assuming the supply to be zero rated), he will have to account for the VAT himself out of the proceeds. If there is any doubt about the VAT position, it is safer to assume the supply is standard rated, charge the appropriate amount of VAT on the invoice and argue about it later.

Time of supply

The *time* at which a supply of goods or services is treated as taking place is important and is called the 'tax point'. VAT must be accounted for to the

Customs and Excise at the end of the accounting period in which this 'tax point' occurs. For the supply of goods which are 'built on site', the 'basic tax point' is the date the goods are made available for the customer's use, whilst for *services* it is normally the date when all work except invoicing is completed.

However, if you issue a tax invoice or receive a payment before this 'basic tax point' then that date becomes a tax point. In the case of contracts providing for stage and retention payments, the tax point is either the date the tax invoice is issued or when payment is received, whichever is the earlier.

All the requirements apply to sub-contractors and main contractors and it should be noted that, when a contractor deducts income tax from a payment to a sub-contractor (because he has no valid CIS6) VAT is payable on the full gross amount *before* taking off the income tax.

Annual accounting

It is possible to account for VAT other than on a specified three month period. Annual accounting provides for nine equal installments to be paid by direct debit with annual return provided with the tenth payment. £300,000.

Cash accounting

If turnover is below a specified limit, currently £600,000, a taxpayer may account for VAT on the basis of cash paid and received. The main advantages are automatic bad debt relief and a deferral of VAT payment where extended credit is given.

Bad debts

Relief is available for debts over 6 months.

Index